Samuel I. Zeveloff

Raccoons
A NATURAL HISTORY

UBC PRESS
VANCOUVER AND TORONTO

IN ASSOCIATION WITH
SMITHSONIAN INSTITUTION PRESS

Drawings by Elizabeth Dewitte

To Barry and Barbara and their beautiful families

Published in Canada by UBC Press, Vancouver/Toronto
www.ubcpress.ca

Editor: E. Anne Bolen
Designer: Janice Wheeler

National Library of Canada Cataloguing-in-Publication Data
Zeveloff, Samuel I., 1950–
    Raccoons

    Co-published by: Smithsonian Institution Press.
    Includes bibliographical references and index.
    ISBN 0-7748-0964-7

    1. Raccoons.   I. Title.
QL737.C26 Z48   2002a   599.76′32   C2001-911724-8

Manufactured in the United States of America
08 07 06 05 04 03 02   5 4 3 2 1

∞ The paper used in this publication meets the minimum requirements of the
American National Standard for Information Sciences—Permanence of Paper for
Printed Library Materials ANSI Z39.48-1984.

C O N T E N T S

Three six-week-old raccoons go out on a limb. Photo by Erwin and Peggy Bauer. The Bruce Coleman Collection.

P R E F A C E

I have written this book to provide a contemporary and thorough treatment of an enormously popular and important North American animal: the raccoon. This book broadly depicts the raccoon's natural history—its evolution, ecology, behavior, and conservation.

The raccoon occupies a unique place in North American culture. As the United States' most economically important furbearer, the raccoon has been eagerly sought by trappers and hunters throughout much of the country, even before its founding as a nation. The history of the raccoon as an object of pursuit and as a ubiquitous denizen of so many of our landscapes has significant cultural aspects that are examined in the pages that follow.

Raccoons have held a special attraction for me since my early childhood. I cannot recall when I first found them to be so appealing, though this event may be related to the broadcast of the Walt Disney television series about the nineteenth-century frontiersman Davy Crockett. Along with many other baby boomers of the 1950s, I eagerly participated in the coonskin cap craze that this series spawned, a fad that is discussed in Chapter 11, "Raccoons and Humans." I recall thinking that those versions of the cap that had a leather or plastic piece on top were simply unacceptable. So my parents indulged my sensibilities and purchased the real item: a hat completely made of fur. That cap left an indelible impression. I can clearly remember the feel of its lustrous fur as well as the striped tail, which on my hat came with a metal snap for easy removal. From that point onward, I was hooked on raccoons.

I had little contact with raccoons until I began working on a master's thesis about their ecology in North Carolina. That was my only extended research experience with this species, yielding findings that are mentioned later. For my doctoral work at the University of Wyoming, I examined patterns of variation in different ecological attributes of mammalian communities throughout North America, analyses that helped me to place my raccoon research within a wider perspective. Several years ago, I organized a symposium on the evolution of the Procyonidae (the raccoon family) at the International Theriological Congress in Mexico to learn about experts' views on this topic. (*Theriology* is the term Europeans often use in place of *mammalogy*, the study of mammals.) Along the way, I never lost my keen interest in and appreciation for this animal. As a zoologist, I have informally continued to monitor its presence in the areas in which I have lived since doing my fieldwork in North Carolina.

Authoring a volume on the raccoon is a logical extension of my interest in writing and editing natural history books. During the past few years, I have written or compiled several books on mammals and wilderness issues, intending for such works to contribute to our understanding of these topics. I hope that broadening the reader's understanding of raccoons will also increase his or her appreciation for all organisms. As is commonly known, an ever-increasing number of Earth's species face uncertain futures because of the numerous environmental changes threatening them, especially the rapid diminishing of their habitats. Ironically, despite the raccoon's success, many of its close relatives are among these threatened species. As is discussed later, the raccoon is successful largely because it is a *generalist*, a species that has an opportunistic lifestyle; *specialists*, organisms with more specific needs, have far dimmer futures.

This book strives to reveal just how unusual the raccoon is. It begins with the history of the raccoon's name and then addresses its origins, describing how it evolved into its present form. The following chapters examine the raccoon's ecology, reproduction, social behavior, and current management issues. The book also has an in-depth discussion of the role that the raccoon has played in human affairs, including a review of its historical and contemporary cultural significance.

To allow the reader to more easily follow the text, few references are cited within it. These are mentioned in the text only when it is essential to do so. All sources are provided in a list at the end of the book. First, those employed in each chapter are listed, followed by the general sources that were used throughout the book. Similarly, I have tried to avoid placing the scientific names of animals and plants in the text. These terms are listed in the appendix.

Ultimately, I hope you come to appreciate the rich life of the raccoon, and the vital association that we have long had with this consummate opportunist.

ACKNOWLEDGMENTS

I am grateful to the many individuals who supported my studies of raccoons and the writing of this book. Phil Doerr of North Carolina State University provided expert guidance as I pursued my raccoon research. I benefited from discussions with the participants of the aforementioned symposium on the evolution of the raccoon family, especially Jon Baskin and Mieczysław Wolsan. Samara Trusso of the International Association of Fish and Wildlife Agencies, Greg Linscombe of the Louisiana Department of Wildlife and Fisheries, and Buddy Baker of the South Carolina Department of Natural Resources provided useful data on the raccoon harvest. Julie Baker, my secretary, has assisted me in multiple ways during the past few years. Various individuals at the Smithsonian Institution Press, including Science Acquisitions Editor Vincent Burke, Assistant Editor Nicole Sloan, Designer Janice Wheeler, and Production Editor Anne Bolen, have all been immensely helpful during this volume's development.

In addition, William Ashworth, of the Linda Hall Library in Kansas City, enhanced my understanding of early European depictions of the raccoon and contributed important historical illustrations. Elizabeth Dewitte drew of all the marvelous new illustrations, such as those of the members of the raccoon family. I trust that the readers will enjoy her work as much as I do. Several other artists and photographers, especially Lynn Kaatz and Julia Sims, kindly allowed me to use their work in this volume.

Last but never least, I am extremely blessed in having such a wonderfully supportive family. My wife, Linda, and our daughters, Abby, Naomi, and Susannah, are always interested in and encouraging of my projects.

An attractive young raccoon relaxes on a log. Photo by Scott Nielsen. The Bruce Coleman Collection.

1

Introducing the Masked Bandit

The North American raccoon is immensely popular. This may be in part because many consider its appearance appealing. A pointed foxlike snout offsets its round face, which is highlighted by a distinctive black bandit's mask across the eyes. The body is typically also round, even plump or pear shaped when in certain positions. Its bushy tail is striped with an interesting series of alternating dark and light bands. Having a face that frequently suggests an air of inquisitiveness or even amusement, the raccoon resembles a chubby, affable clown rather than the considerably strong and occasionally ferocious animal that it is.

Surely, its handsome, thick fur coat is also responsible for its allure. The lustrous fur is usually a dark brown, heavily flecked with gray, black, and blond hairs. As is typical for mammals, it is lighter across the belly. However, as is common for mammals with broad distributions, the raccoon exhibits considerable variability in its coat color and pattern. (Its physical characteristics and numerous subspecies are described in detail in Chapters 4 and 5, "Form and Function" and "Distribution and Subspecies," respectively.)

The raccoon's charm is most certainly also based on its considerable repertoire of amusing behaviors. It engages in seemingly unusual acts for a wild animal, the most well known of which is its supposed washing of food prior to eating it. Its curiosity is almost as legendary as a cat's, and its remarkable dexterity enables it to adeptly manipulate and pry open various items. (These traits are reviewed in detail in the following chapters.)

However, not all of the raccoon's press notices are positive. Although young raccoons may make enchanting pets, when they grow up their seemingly insatiable curiosity, general lack of responsiveness to training, and thus untrustworthiness usually will try the patience of even the most devoted owner. As an *omnivore*, an animal that feasts upon a wide array of foods, the raccoon often upsets the interests of many other consumers, especially humans. It is a notorious pest of crops such as corn and it is a well-known thief of eggs from bird nests and nest boxes erected to enhance waterfowl reproduction. Expanding their range into suburbia, raccoons have become nuisances in other ways, using their dexterity to open secured garbage cans and then wreaking havoc with their contents. Unfortunately, they have also become a public health concern as a significant carrier of rabies in an expanding area of North America.

How the raccoon was given its common as well as its scientific name, *Procyon lotor*, involves not only its appearance and behavior but also interesting historical events, such as early European associations with Native Americans and the development of *taxonomy*, the science of classifying organisms. The word *raccoon* is derived from the language of the Algonquins, a group of Native American tribes. The term *Algonquian* is an adjective used to describe their language group; this term has also been used to describe the people who speak these languages. Algonquian is one of the most well-known Native American language groups and might have been the language of the first natives the English colonists met. Many Algonquian words are still part of the English now spoken in North America, though of course quite a few have changed through time.

Various sources recount that seventeenth-century English settlers living in what became the eastern United States based their word *raccoon* on the Algonquian word *arakun*. It is the shortened form of *arakunem*, which roughly translates as "he who scratches with his hands." Algonquins such as Chief Powhatan and his famous daughter Pocahontas are believed to have pronounced it as *ah-rah-koon-em*. The word *arakun* has been transcribed in other forms such as *aroughcan, aroughcun,* and *arathkone*. The names of several other mammals are also based on Algonquian terms, including the skunk, woodchuck, and chipmunk.

Although raccoon has become the common name for the species, a few other forms continue in colloquial usage. The shortened form *coon* is often used, especially when describing clothing made from raccoon fur, such as a coonskin cap. *Ringtail* is used very rarely as a common name for the raccoon and is usually reserved for one of its relatives, *Bassariscus astutus*.

Many other names have been bestowed upon the raccoon, though clearly none have worn as well as the Algonquian word reflecting its supposed scratching behavior. For example, at about the time when the first raccoon furs arrived in Germany, which could have been as early as the sixteenth century, they were given the name *schupp*. This word had been previously used to describe various related items: a fish vendor, a fisherman, fish scales, and even the person who scraped off the scales. Perhaps the German furriers gave the raccoon this name in reference to its fishing skills, though they also used it for the marten, a member of the weasel family, Mustelidae. Variations of this term, such as its Dutch version, *schob*, spread throughout Europe.

As is discussed in Chapter 11, when the seventeenth-century French settlers in northern North America used the term *chat sauvage* or "wild cat," they were likely referring to the raccoon. They also called it *le raton laveur*, or "the rat that washes." In the early eighteenth century, in what is now the northeastern United States, Dutch settlers called the raccoon a *hespan* and the Swedes referred to it as an *espan*. The raccoon's present German name, *wachsbar*, means "little bear." Somewhat similarly, its Mexican name, *osito lavador*, is defined as "little bear washer." Indeed, the raccoon has a great variety of names. A comprehensive listing of these is provided in the book *Raccoons in Folklore, History and Today's Backyards* by Virginia Holmgren.

European naturalists named and described many North American animals following their sixteenth-century discovery by European explorers. Holmgren reports that the first published portrait of a raccoon appeared in the 1551 book *Historia Animalium* by Conrad Gesner, a noted Swiss naturalist. (Incidentally, Gesner also published the first illustrations of fossils in 1565.) Holmgren is slightly mistaken; this portrait of a so-called raccoon, entitled *Mus Indicus*, did not appear in the first edition of *Historia Animalium*, though it was printed in later ones, perhaps as early as 1554. It was also subsequently printed in a volume called the *Thierbuch*, an abridged 1563 German translation of *Historia Animalium*. Gesner placed this animal in a now-defunct group that was designated by the Latin word *mus*, which included mammals with pointed snouts such as the sable. This term now refers to another pointy-snouted group, the genus of mice such as the ubiquitous house mouse, *Mus musculus*.

According to William Ashworth, a leading authority on the scientific illustrations from that period, Gesner's *Mus Indicus* may not have been a raccoon at all. Instead, his portrait may have depicted a coati, a close

Illustration of *Mus Indicus* from Conrad Gesner's *Thierbuch* (Zurich, 1563, leaf 115v), an abridged German translation of his Latin *Historia Animalium*. This illustration was hand colored when the book was published. From the collection of the Linda Hall Library, Kansas City, Missouri.

cousin of the raccoon. Ashworth believes that the earliest European illustration of the raccoon appeared almost 100 years later in 1655 in a volume entitled *Museum Wormianum* by Ole Worm. Ironically, Worm describes this animal as a "coati [racoon]" [*sic*]. Ole Worm maintained a private collection of specimens, a "cabinet of curiosities," which was later transferred to the Danish royal collection in Copenhagen.

The raccoon's earliest scientific name seems to have been *Vulpi affinis Americana*. This appeared in a 1693 publication by the English naturalist John Ray. The term *Vulpi* is similar to *Vulpes,* the present genus name of the red fox. Foxes are members of the Family Canidae, which includes wolves and dogs. Thus, based on the name's second part, Ray apparently believed that the raccoon had an affinity to the fox. As mentioned in the following chapters, many early observers also mistakenly believed that the raccoon was closely related to the canids. The last term, *Americana,* would now be considered a subspecies name, although it would not be capitalized. Ray's designation for the raccoon was apparently used for its various forms in North, Central, and South America. Today, several distinct raccoon species are recognized, each of which is reviewed in a subsequent chapter.

The great eighteenth-century Swedish naturalist Carolus Linnaeus devised our present scientific system of naming organisms. In his arrangement, an organism's name has two parts: its genus or generic name and the species name or specific epithet. The subspecies name, which refers to a geographic race or variety, may then follow. Although the technical name Ray used seems to fall within these guidelines, it was Linnaeus who

Illustration labeled "coati [racoon]" [*sic*] in Ole Worm's *Museum Wormianum* (Leiden, 1655, p. 319). From the collection of the Linda Hall Library, Kansas City, Missouri.

fully developed and refined this system. In periodic revisions of his monumental work, *Systema Naturae,* he tried to provide scientific names for all of Earth's living forms, basing them on what was known about species' relationships. For example, those that he considered to be closely related were grouped together in a common genus.

By the time Linnaeus published his second edition of *Systema Naturae* in 1740, he had apparently thought about the resemblance between raccoons and bears. He named the raccoon *Ursus cauda elongata,* distinguishing it from his true bear, *Ursus cauda abrupta.* Obviously, he regarded the raccoon as a long-tailed (*cauda elongata*) cousin of a short- or abrupt-tailed (*cauda abrupta*) bear. Just as early naturalists thought that the raccoon was a close cousin of the canids, many later generations of biologists have considered it to be closely related to the bears. However, the latter assumption may also be wrong. In a 1747 publication, Linnaeus provided an illustration and a lengthy description of the raccoon. This work includes what may be the first picture of its unusual s-shaped penis bone, its *os penis* or *baculum,* the plural form of which is *bacula.*

In 1753, one of Linnaeus's students, another Swede named Peter (or Pehr) Kalm, described the raccoon during his travels in Pennsylvania and New Jersey. He remarked on how easily they were tamed, and noted that Native Americans used the raccoon's penis bone as a pipe-cleaning tool. Kalm may have told his mentor about the raccoon's supposed habit of washing its food because in his tenth and definitive edition of *Systema Naturae* published in 1758, Linnaeus changed the raccoon's name to *Ursus Lotor* (in modern usage the last term would not be capitalized). His continued use of the term *Ursus* reveals that he still considered raccoons and bears to be close cousins. In changing its species name to *Lotor,* meaning "washer," Linnaeus was referring to a behavior that is more accurately described otherwise.

Soon thereafter in 1780, a German natural history professor with the lengthy name of Gottlieb Conrad Christian Storr decided to place the raccoon in its own genus, which he termed *Procyon*. Because its full scientific name was based on Linnaeus's *Ursus Lotor,* the raccoon became classified as *Procyon lotor,* and this has been its scientific name ever since. The raccoon's full name is actually *Procyon lotor* (Linnaeus, 1758). Even though Storr proposed the now-accepted genus name, the information in parentheses indicates that Linnaeus is still the authority for its full name because he was the first to introduce a portion of it.

Procyon is a Latin term meaning "before the dog" or "early dog." Why Storr selected this term is not known, though it is highly unlikely that he thought of raccoons as being ancestral to dogs. He chose this term before much if anything was known about evolutionary processes. Charles Darwin, the founder of modern evolutionary theory, was not even born until 1809, 29 years after Storr proposed his name for the raccoon. The Latin prefix *pro* (which is originally a Greek term) could have several meanings, including "before," "for," "just as," or "like." Thus, Storr might have been referring to the raccoon's doglike ways, or perhaps he was just suggesting that it was different enough from the bears to merit its own genus. In the previous century, astronomers had also used the name Procyon for a newly discovered binary, or double, star in the constellation Canis Minor, which appears in the heavens shortly before Sirius, the Dog Star. In any case, we now know that the raccoon's ancestors appear to have descended from members of the dog branch of the Order Carnivora by about 25 million years ago. These origins and relationships are fully explored in the following chapter.

Despite the passing of almost 20 years since Storr changed the raccoon's genus, in 1798 the illustrious French zoologist Georges Cuvier still described South America's crab-eating raccoon as *Ursus cancrivorus,* retaining the bear's genus name that Linnaeus had given to it. This choice might suggest that Cuvier disagreed with Storr's new name and its implications, though that is not clear. Ultimately, though, the crab-eating raccoon became classified as *Procyon cancrivorus,* adopting the genus name that Storr had proposed for its North American cousin. Through time, several other species and subspecies designations have been used for the various types of raccoons. Some of the names that had been assigned to certain raccoons are now considered to be invalid based upon issues such as deviations from internationally agreed upon taxonomic rules as well as disagreements about whether some raccoons are distinct species.

As mentioned previously, the second half of the raccoon's scientific name, *lotor*, means "washer." Undoubtedly, they may wash crayfish and the other items they catch in or near water to clean off the sand and dirt from them. But most of the time that raccoons are putting their food into water, they are merely dousing or dunking it, an activity that has nothing to do with cleanliness. Clean, dirty, wet, and dry foods are all dunked with similar frequency. Raccoons probably handle their food in water to provide greater tactile sensation. As is discussed in Chapter 4, their hands have highly developed nerves that correspond to specific areas of the brain. Obviously, they use their hands to search for food. Yet in certain circumstances such as in shallow water, their vision and sense of smell may be limited. Because their forepaw skin becomes softer and more pliable when wet, the sensitivity of their paws should be increased while they are in water.

Raccoon food dousing might also be what is called a fixed motor pattern. Because raccoons use a dunking motion when searching for aquatic prey, "washing" their food under other circumstances may simply be an imitation of this movement, even when they are not foraging. Much of the time, raccoons eat far from water where washing food is not even an option. Previously, raccoons were thought to lack salivary glands and thus would have needed to moisten their food. They are now known to have well-developed salivary glands that function the same as in other animals: to aid in swallowing and in the production of digestive enzymes.

Therefore, oddly enough, both parts of the raccoon's scientific name, *Procyon* and *lotor*, are based on terms that do not accurately apply to its existence. Such names, however, are difficult to alter once they have been accepted. Individuals must go through a lengthy process to justify what they believe to be a more apt name for an organism, and altering one with a long-term historical basis could be resisted. Thus, although the raccoon is not an ancestor to the dog and despite the fact that it is not much of a washer, it will likely be classified as *Procyon lotor* far into the future.

2

Raccoon
Origins

To better understand the raccoon's evolutionary history, one must first consider the larger groups to which it belongs. This exploration begins with a brief look at the mammals, the major animal group that includes the raccoon. The mammals are in the Class Mammalia, one of several classes of *vertebrates,* or animals with backbones. Like other taxonomic entities, this class is composed of members that share certain characteristics. These include having fur or hair at some point in their lives and females having the capability of *lactation,* the production and secretion of milk. A lesser-known trait, but one that characterizes all mammals, is the possession of a muscular diaphragm that separates their pulmonary and abdominal cavities.

The mammals are divided into three major groups: the monotremes, the marsupials, and the placentals. Confined to Australia and its surrounding region, the monotremes are a primitive group of only three species that have retained several reptilian traits, such as egg laying. These are the platypus and two different types of spiny anteaters, or echidnas, one of which is located in Australia and the other in New Guinea.

The marsupials are a considerably larger group, consisting of about 250 species. They include well-known animals, such as kangaroos, as well as species with exotic names like the noolbenger. Marsupials produce young that are virtually embryonic at birth and almost always complete their development in a *marsupium* or pouch. Primarily distributed throughout Australia, a fair number of marsupials are also in South America. There is only one North American representative, the opossum, which is closely linked to the raccoon in the folklore of the southeastern United States.

The placentals are by far the most numerous, widely distributed, and thus best-known group of mammals. About 3,800 species of placentals exist, including such diverse types as rodents, bats, and whales. Major groupings of such types compose *orders*; rodents, for example, are in the Order Rodentia. Placental mammals have relatively long pregnancies that allow them to produce comparatively well-developed offspring. Even so, the developmental stages of their newborn vary considerably. Their lengthier gestations are facilitated by the structure that gives the group its name, the *placenta*. The placenta, a thick organ that develops during pregnancy, is positioned between the inside uterine wall and the fetus. It has a rich meshwork of blood vessels that supplies the developing embryo with nutrients from its mother.

CARNIVORE EVOLUTION

Raccoons belong to the Order Carnivora, a subdivision of the placental mammals. It is a comparatively small order of about 246 species. *Carnivora*, which literally means "eaters of flesh," obviously refers to the meat-eating habits of the majority of its members. However, some of its species, including this book's subject, have rather broad diets and are better categorized as *omnivores*. A few members of this group do not even eat meat, such as the African aardwolf, which largely feeds on termites. Thus, rather than always refer to the species in this order as carnivores, it is helpful to distinguish them from the many unrelated meat-eating species by capitalizing the "c" to create the term *Carnivores*, as suggested by David Macdonald in his book *The Velvet Claw*.

The species in this order are frequently recognized by their large, pointed canine teeth, which are primarily used to kill prey. An even more distinguishing dental trait is their set of specialized shearing teeth known as *carnassials*. In adult Carnivores, these teeth are the last upper premolars and first lower molars on each side of the jaw. Usually bladelike, they work together so that meat is sheared as the animal chews it. Each carnassial has two narrow cusps separated by a notch that holds the meat while it is being cut. Even newborns have carnassial-like baby teeth, though these are the third upper and last lower premolars. As most of them begin to eat meat before they are weaned, these baby carnassials can be critical to their survival.

The development of the carnassials is considered to be a pivotal event in Carnivore evolution. Furthermore, because these teeth have persisted throughout the evolution of this order, they have been useful in revealing its history. Coated with enamel, the body's hardest tissue, teeth are the most enduring of all fossil materials. Because the distinctive carnassials occur sufficiently in the fossil record, the history of this order has been well documented. The earliest known carnassials date from about 65 million years ago, at the dawn of what is known as the Cenozoic era, though they may have occurred several million years earlier. This is the time when dinosaurs became extinct and the mammals began their ascendancy as the dominant group of terrestrial vertebrates. The immense variety of ecological niches the dinosaurs once occupied, particularly those of the large herbivores and carnivores, became available to the mammals. (Note that the dates of the geological time periods mentioned herein are approximate and can vary somewhat depending on the source. Those used here are mostly from Michael Woodburne's volume *Cenozoic Mammals of North America: Geochronology and Biostratigraphy.*)

Interestingly, the earliest mammals to become large predators were not placentals; they evolved from a kind of marsupial that occurred in South America and Australia. The first placental mammals to become habitual meat eaters were not even from the Order Carnivora. Ironically, they arose from a group of vegetarians, an extinct order called the Condylarthra. Initially occurring in the northern hemisphere, condylarths are the ancestors of all existing large placental herbivores, including such different species as the African elephant and the moose. Approximately 60 million years ago, some of the early condylarths were carnivorous. One intriguing group, the bear-dogs of the Family Arctocyonidae, consisted of species resembling a cross between these animals, though they are not related to either of the existing groups. Nor are they related to a later, also extinct group—the giant bear-dogs of the Family Amphicyonidae. The bear-dogs ultimately gave rise to a group of carnivores called mesonychids (the Family Mesonychidae). They became the dominant terrestrial predators for an enormous length of time, about 20 million years, primarily during the Eocene epoch, which lasted from about 38 to 58 million years ago. One North American form, the wolflike *Mesonyx*, had a powerful skull with molars that may have been capable of shearing. Yet like the other condylarths, it had hooves rather than claws like the modern Carnivores.

Today's Carnivores, on the other hand, likely arose from creatures similar to one called *Cimolestes*, which evolved from a different lineage than the condylarths. *Cimolestes* is the genus name of several extinct squirrel-sized mammals that lived a little more than 65 million years ago. They likely preyed on various invertebrates and perhaps small vertebrates as well. Their cheek teeth were bladelike and thus could have been capable of carnassial-like slicing. By about 58 million years ago, two lineages emerged that utilized this flesh-shearing action, the Carnivora and the Creodonta.

The first clearly distinguishable species of the Carnivora were the miacids or miacoids, members of the Family Miacidae. Mostly small and thin, they varied from being the size of a weasel to that of a marten and possessed enlarged canines and shearing carnassials. The Miacidae was the sole family in this order for about 20 million years. Although considerable guesswork is involved in reconstructing their natural history, the miacids have been characterized as tree-dwelling hunters that lived in lush forests.

Nevertheless, it was the other early carnivorous line, the Creodonta, that emerged as the first truly dominant meat eaters. They also flourished from about 35 to 55 million years ago and included a variety of species similar to modern dogs, bears, hyenas, and cats. However, unlike the miacids, the creodonts are not considered to be on the main line of Carnivore evolution. The group that evolved from the miacids into today's Carnivores only finally assumed dominance over the creodonts about 20 to 30 million years ago. Some researchers have suggested that the Carnivores ultimately succeeded because they were smarter than the creodonts. This opinion, however, is not supported by a comparison of their brain sizes. A more significant difference between these groups seems to be which of their teeth functioned as carnassials. Those of the creodonts were located further back in the mouth, at their second and third upper molars and at the third and fourth lower molars. Because they lacked teeth behind their carnassials to grind fibrous plant material, they were apparently mostly restricted to meat eating. The Carnivores, on the other hand, had several teeth behind their carnassials that could eventually become modified for vegetation grinding. Thus, they may have been better positioned to radiate into species with varied diets.

Slightly more than 30 million years ago during the Oligocene epoch, which lasted from about 24 to 38 million years ago, global climatic changes are believed to have occurred that resulted in more pronounced seasons. Such shifts may have yielded periods of abundance of fruit and insects, which the potentially versatile Carnivores would have been able to exploit effectively. Ironically, this capacity to diverge from meat eating may have led to their dominance over the creodonts. Though this explanation might seem simplistic and certainly cannot be tested, it reveals how an overspecialized lifestyle can lead to a species' demise.

DIVERSIFICATION OF THE CARNIVORA

Hence, the miacids and their descendants ultimately gave rise to today's successful Carnivores. At least 55 million years ago, the early tree-dwelling miacids apparently divided into the two main branches of the Carnivora: the vulpavines and viverravines. Several variants of these names exist, but these should suffice for this discussion. Initially, the vulpavines were a New World group, occurring in what became North America. Its species, many of which likely resembled today's martens, evolved to become the dog branch of the order, a suborder called the Caniformia. It has also been referred to as the Superfamily Canoidea or Arctoidea. (Subsequently in this chapter, the latter name is more appropriately used to describe a group within the dog branch.) The viverravines, alternately, were at first an Old World or Eurasian group. It gave rise to the other modern Carnivora suborder, the Feliformia or cat branch, a group also referred to as the Superfamily Feloidea or Aeluroidea. Early viverravines were similar in appearance to modern genets, small catlike creatures in the Family Viverridae.

Although evidence for the early evolution of this order is sketchy, by about 40 million years ago, the early miacids had evidently undergone a relatively rapid diversification that led to today's Carnivores. Many of the modern families appeared between 35 to 40 million years ago. Those in the dog branch that emerged at this time are the Canidae (dogs and their relatives), the Ursidae (bears), the Mustelidae (weasels and their relatives), as well as the now-extinct Amphicyonidae (giant bear-dogs). The Procyonidae (raccoons and their relatives) evolved later, by about 25 million years ago in the late Oligocene.

In the cat branch, the Felidae (cats and their relatives) and the Viverridae (genets and civets) appeared at around the same time as the early members of the dog branch. The youngest terrestrial Carnivore family is the Hyaenidae (hyenas and the aardwolf). It evolved from the Viverridae about 15 million years ago in the middle of the Miocene epoch, which occurred from about 5 to 24 million years ago. By around 35 million years ago, the geographical separation of the Old World and New World Carnivores had ceased. The Bering Strait, a land bridge between Asia and North America, had opened, permitting species from each group to ven-

The evolution of the Order Carnivora, focusing on the dog branch. Beginning dates of the geological epochs are: Paleocene—66.5 million years ago; Eocene—58 million years ago; Oligocene—38 million years ago; Miocene—24 million years ago; Pliocene—5 million years ago; Pleistocene—1.8 million years ago; and Recent—11,000 years ago. The raccoon's family, the Procyonidae, arose about 25 million years ago. Illustration by Elizabeth Dewitte; based upon various references in Chapter 2 and especially Macdonald 1992.

ture into the other's domain. Finally, by about 7 million years ago, all of the modern Carnivore families were present in both North America and Eurasia.

Obviously, mammalian teeth contain invaluable information about this group's history. Of course, they can also reveal a great deal about an animal's feeding adaptations. The teeth of dogs and cats exemplify the differences in their branches' dentitions. For example, the members of the cat family are highly specialized for meat eating. Other than the carnassials, they have lost all of their lower molars. Their only upper molars are two tiny bladelike teeth, each of which is located behind an upper carnassial. The back teeth are largely used to slice meat as they lack the crushing surfaces common to most molars. The result is that strips of meat are only slightly digested in the mouth before being passed to the stomach. The felids' *hypercarnivory*, the virtually exclusive eating of meat, is not typical of every member of their suborder; many eat a wider range of foods.

The carnassials of the species in the dog family appear to be the least changed from those of their ancestors. The lower ones have a broad shelf at the back that occludes with the upper first molars. These lower carnassials not only shear food at their front but can also crush it at their rear. Like various other mammals, canids thus use their molars for chewing, or mixing food with saliva so that digestion can begin in the mouth. This allows them to process various foods, including meat, bone, invertebrates, and plants. Unquestionably, this capability has great survival value: the wider the food range, the greater the ability to shift one's diet if environmental conditions should change. The raccoon and its cousins are among those species in the dog branch that benefit from such feeding versatility.

Since their formation, each of the two major branches of the Carnivora has continued to diversify, producing a variety of now-extinct and modern families. The cat branch presently also includes the Herpestidae (the mongooses), though some experts feel that it is actually a subfamily of the Viverridae. Most authorities also include the *pinniped* marine Carnivores in the dog branch: the seals of the Phocidae, the sea lions of the Otariidae, and the walrus of the Odobenidae. Some, however, continue to separate the Carnivora into terrestrial and marine divisions, whereas others even place these marine Carnivores in their own order.

THE DOG BRANCH

In the last several years, researchers have used a broad array of molecular techniques to study the relationships among species as well as their evolutionary histories. Such research often supplements the understanding gained from fossils, though it occasionally results in different interpretations of the connections among species or their histories. These analyses have generated some provocative findings about the history of the Carnivores, including that of the raccoon family. They suggest that the most ancient division in the dog branch, that between the canids and their close relatives, might have occurred about 50 million years ago, with several other dog and cat lineages appearing about 10 million years later. Using DNA analyses, the raccoon family has also been estimated to have separated from the others in the dog branch about 40 million years ago. Similar evidence has indicated that the bear and raccoon families diverged from a common ancestor between 30 to 40 million years ago.

Obviously, such estimations have resulted in considerably earlier dates for these events than those projected from the fossil record. The fossil evidence, as mentioned, suggests that the raccoon family emerged from its ancestors about 25 million years ago, much later than the date furnished by the molecular studies. As might be expected, disagreements exist about the validity and interpretation of the different dating methods. In addition, the suggestion that raccoons and bears diverged from a common ancestor differs from a recent understanding that the procyonids are actually more closely related to the weasel family, which is discussed in detail in the following section.

The raccoon and the other members of the dog branch probably descended from a species resembling *Vulpavus*, a genus of tree-top hunters that lived about 40 million years ago in North America. The families that descended from this *Vulpavus*-like ancestor are distinguished by their number of molars: the canids have three upper and three lower molars; bears have two upper and three lower molars; weasels and their kin have one upper and two lower molars; whereas raccoons and their relatives have two upper and two lower molars. In the line leading to the latter three families, the last upper premolars were lost at least 35 million years ago, and the members of the latter two families eventually also lost their last lower molars. Normally, losing all of these teeth would reflect an emphasis on meat eating. However, the significance of such tooth loss in the early bears and procyonids is uncertain because their cheek teeth had developed more of a crushing function soon after their families appeared. Their shearing cusps became blunted and the grinding surfaces had enlarged, changes that were associated with their emerging omnivorous and herbivorous habits. Today, the carnassials of the species in these families are the least developed for meat eating among the Carnivora; this style of feeding is called *hypocarnivory*.

The true dogs or canids remained in North America for millions of years. Another family in the dog branch, the aforementioned giant bear-dogs, expanded their distribution by migrating to what were then lush forests in Eurasia. While these ground-dwelling omnivores initially flourished, later they may have encountered competition on several fronts: from canids in the New World, doglike hyenas in the Old World, and bears and procyonids in both regions. About 15 million years ago, these bear-dogs declined, perhaps because they could not successfully compete with either the heavier bears or the faster dogs. Most significantly for this discussion, their decline may have benefited the procyonids by opening up additional niches for them.

The giant bear-dogs might have persisted had their diet shifted even more toward vegetation. The raccoon and bear families apparently had already incorporated much vegetation into their diets. As indicated, most of the families in this branch began to split from the main canid line about 35 million years ago. Until then, their predecessors had been mostly carnivorous, but these newer families began to increasingly consume plants. The bears and the smaller procyonids may have sustained themselves primarily on fruits and nuts, and the early members of both families might have increased their range of foods by foraging in the trees as well as on the ground. Unfortunately, these scenarios are only barely supported by a spotty fossil record.

ORIGINS OF THE PROCYONIDAE

Even after a careful study of the available fossils, interpreting a group's history is rarely easy. For example, early attempts to understand the evolution of the raccoon family resulted in its members being erroneously classified as descendants of a subfamily within the dog family, the Cynarctinae. Their fossils' similarities appear to be due to both groups switching from a carnivorous to an omnivorous life style. Their likeness thus seems to have been caused by *convergent evolution*, a process wherein unrelated species develop similar traits by becoming adapted to similar niches.

Now the raccoon family is firmly established as belonging to a group called the arctoids. Formally known as the Infraorder Arctoidea, this group also includes the weasel family, bears, the aforementioned marine Carnivores, and the extinct giant bear-dogs. Initially, the arctoids occurred in the northern parts of Eurasia and North America. They were probably a distinct group by about 38 million years ago because their common distinguishing traits had become especially apparent by then. These consist of certain characteristics at the base of the skull, or their *basicranial* features. Arctoids dominated the mammalian faunas of this northern region for a long time, probably from soon after they appeared until the Recent epoch, which began about 11 thousand years ago. Before the raccoon family evolved, their *adaptive zone,* the environmental conditions in which a group is most successful, was mostly occupied by other small bearlike arctoids. These included the European *Cephalogale,* which is considered to be the first bear genus, and various North American species.

Yet despite sharing some dental features, such as having additional molar cusps, and their possible linkage through molecular evidence, the raccoon and bear families are apparently not as closely related as has long been assumed. Leading paleontologists and anatomists have convincingly argued that the history of the procyonids is most closely connected with that of the weasel family, Mustelidae. This affiliation is based on certain structural similarities that the two families share, including dental traits and characteristics of their skulls' basal areas. Accordingly, they have been united in a group called the musteloids.

The musteloids first appeared in North America between 33.4 and 37.1 million years ago. Their principal fossils from this period are from the genus *Mustelavus.* In Europe, these animals existed later, from 29.4 to 33.4 million years ago, and were called *Mustelictis.* Both are apparently within the stem group from which the other musteloids developed. However, this is not certain; *Mustelavus* has alternatively been considered to be an early member of the raccoon family. The musteloids may also include the seal family, though this view has been disputed as well. Despite the disagreement over these issues, a consensus is growing that the raccoons and their relatives should be combined in a larger group with the weasel family rather than with the bears.

One feature that connects the raccoon and weasel families is that the ancient members of each group share the same dental formula; they both have the same number of specific types of teeth. Furthermore, as stated, they also have similar elements at their skulls' bases, in particular, features in their auditory regions. An intriguing structure called the *suprameatal fossa,* a deep hollow in the middle-ear cavity, has been used to establish their relationship. Oddly, its function, if any, is a mystery. Yet because its form is so similar and has such a complex shape in procyonids and certain mustelids, this implies that these families developed from a close common ancestor. Today, however, this structure is different in the two groups. It has been argued that the fossa in the raccoon family more closely resembles its original form and that the one in the weasel family was derived from it. Other views suggest that this structure evolved independently in each group or that the one in the weasel family is the more primitive type.

Although this structure occurs in many other arctoid mammals, it does not develop beyond a preliminary stage in the bears or in the red or lesser panda. Unlike those of the other Carnivores, the suprameatal fossa of the procyonids is conspicuously well developed. It is both deep and extensive, and it opens broadly where it faces downward. Among present species, it only appears this way in the raccoon family. Though some extinct and living procyonids have such depressions that deviate from this pattern, these occurrences seem to be *secondary developments*, meaning that they evolved from the initial version.

Despite these connections between the raccoon and weasel families, some researchers remain unconvinced about their affinity. Some argue that these animals share relatively few traits other than the suprameatal fossa that are derived from a common ancestor. Early procyonids and some bears from the same period also possess very similar teeth and share certain skull features in the basal area, suggesting that the raccoon family may be more closely linked with the bear family after all.

In either case, the raccoon family finally emerged as a distinct entity in Eurasia toward the end of the Oligocene. Fossils found in France and Germany reveal that a diverse group of apparent procyonids lived there from about 25 million years ago until the end of the early Miocene, about 18 million years ago. This group included forms such as *Amphictis*, which appeared early in this period, and *Broiliana* and *Stromeriella*, which occurred near its end. Another fossil from this time, that of *Sivanasua*, may be evidence of the earliest known ancestor of the red panda. To classify these animals based on their skulls is often difficult. For example, *Plesictis* has been regarded both as a significant early procyonid and as a member of the weasel family. Thus, the evolution of the modern raccoon from this genus is doubtful.

Amphictis was a rather successful genus that included several fairly different species. It is the one that most experts agree included the earliest known procyonid. As explained, classifying either the nearly concurrent *Plesictis* and the earlier *Mustelavus* as the raccoon family's first member raises concerns. *Amphictis* retained its second lower and second upper molars and lost its third lower molar, all of which are key procyonid traits. However, it only had a rudimentary suprameatal fossa, though this structure may have become well developed in its later species. As a result, even the position of *Amphictis* is uncertain; it has also been placed near the stem of both the raccoon and weasel families. Yet another, even earlier European mammal, *Pseudobassaris*, which occurred about 29 to 30 million years ago, has also been mentioned as the earliest true procyonid, though many authorities do not believe that the family existed then.

The first procyonids in the New World most likely evolved from just a few of these early western European forms. Although *Stromeriella* is the only one for which confirmed fossil evidence in North America exists, other procyonids could have also made the journey over. They probably first appeared here as early as 19 million years ago, though they may have arrived later, between 16 and 19 million years ago. These stem procyonids ostensibly migrated across the Bering Strait that connected Asia and Alaska, from which their descendants would have traveled inward throughout North America. Apparently all of today's North and South American procyonids descended from just a few Eurasian species, which helps to explain their genetic similarity. However, no fossil evidence exists that directly supports this explanation.

DIVERSIFICATION WITHIN THE PROCYONIDAE

Within the raccoon family are smaller groups of related species known as subfamilies or *clades*. Although which species tend to be most closely related is generally understood, the number and composition of these groups are still highly subject to interpretation.

The first of these subfamilies is the Procyoninae; its members are called procyonines. These species are believed to have diverged the most from the original type of procyonid. The aforementioned fossil *Broiliana* is regarded as its most primitive member. Some experts include the raccoons, coatis, and ringtails in this subfamily, though the raccoons and coatis form an even more closely related assemblage within it.

Given this affinity between coatis and raccoons, some authorities have placed the ringtails and their close relatives in a different subfamily: the Bassariscinae or bassariscines. Yet another subfamily arrangement that includes just some of these bassariscines is the Potosinae. It consists of the olingos and the kinkajou. But because the kinkajou is so unique, it has alternately been classified as the sole member of this subfamily. To complicate matters even further, some authorities do not consider either of these subfamilies to be legitimate but instead assign their species to the originally mentioned subfamily, the Procyoninae.

Then there is the Old World red panda. If it indeed is a procyonid, it would be placed with its extinct relatives in the Subfamily Ailurinae. It appears to be closely related to the species in an extinct subfamily, the Simocyoninae. The mammals in these groups are similar enough to each

Recent	
Pleistocene	Red Panda Kinkajou Olingos Coatis Raccoons Ringtails
	Ailurus
Pliocene	*Parailurus Potos*
	Nasua Procyon
	Bassaricyon
	Cyonasua
	First S. American Procyonid
	Paranasua
	Bassariscus
	First New World Procyonid (N. American)
	Amphictis
Miocene	
	Procyonidae
Oligocene	

The evolution of the Family Procyonidae. Epoch dates are given in the previous figure caption. Because fossil evidence is scarce, the dates when certain groups developed and the nature of these groups' relationships can only be hypothesized. Illustration by Elizabeth Dewitte; based upon various references for Chapter 2.

other (for example, they both possess enlarged second molars) yet differ sufficiently from the other procyonids that combining them into their own distinct family, the Ailuridae, might also be reasonable. *Amphictis*, the reputed first procyonid, may be closely related to the species in one of these groups.

The Simocyoninae was the principal group of procyonids that remained in Eurasia after the others had migrated to North America. Their cousins, the Ailurinae, apparently persisted there as well. Some simocyonines had wide distributions, occurring in Europe, Asia, and North America. At least one of them, *Simocyon*, was extremely widespread. It ranged from Spain to China and was even successful in western North America. Its powerful jaws and well-developed carnassials suggest that it had a hyenalike style of predation. Yet it may have suffered from competition with other Carnivores, such as *Osteoborus*, a short-faced hyenalike canid. The simocyonines probably emigrated here from Asia across a Bering Sea connection in the middle or late Miocene, just as the stem procyonids appear to have done so earlier. However, they were not able to successfully establish themselves here. Both the simocyonines and ailurines are known only from rare, isolated specimens in North America. By about 5 million years ago, the simocyonines had become extinct.

In contrast, the major radiation of the raccoon's subfamily, the Procyoninae, took place in the New World. Though several occurred in North America, their fossil record suggests that they reached their greatest diversity in the Neotropics or New World Tropics, a phenomenon that still holds true today. Based on their dentition, none relied strongly on meat eating. An advanced, widespread species of one of them, *Arctonasua*, was rather large, likely weighing 30 to 40 kg. Its teeth had bulbous cusps, suggesting that it engaged in a bearlike omnivory. On the other hand, the teeth and auditory regions of both fossil and present-day bassariscines (ringtails and their relatives) appear to be similar to those of the earliest procyonids. The ringtail, for example, has a hypercarnivorous dentition that strongly reflects its meat-eating habits. Given that they have changed the least from their ancestors, some of today's bassariscines may be more directly related to the family members that first appeared in North America 16 to 19 million years ago.

During both the early and middle Miocene, these ancient bassariscines occurred in what is now the Great Plains of North America. They included *Probassariscus antiquus*, which appears to have been the family's first new member on the continent about 19 million years ago. Representatives of this genus persisted for at least another 7 million years and perhaps considerably longer. About 17 to 18 million years ago, several other procyonids lived in that area as well. They included *Stromeriella* (which, as mentioned, is the only North American fossil from one of the ancestral European forms) or perhaps a close relative of it. Yet another, the coatilike *Edaphocyon*, was also discovered there. The oldest North American members of the procyonine subfamily belonged to this genus, though at least one of them could have been a Eurasian emigrant. Thus, the raccoon family was apparently rather diverse on the Great Plains at that time. Presumably, this area was well vegetated and threaded with watercourses. Subsequently, much of family's fossil record is from Texas and Florida, especially that from the late Miocene, which ended some 5 million years ago.

Curiously, many of the prehistoric procyonids were adapted for a somewhat *fossorial*, or digging, lifestyle. The remains of one early Miocene species, *Zodiolestes daimonelixensis* (which has also been placed in the weasel family), were discovered curled up in the home of a burrowing beaver. While it is conceivable that *Zodiolestes* hunted these ancient rodents, it is difficult to draw conclusions about predator-prey relationships from fossils. After what was apparently a short trial with such habits, most of the North American procyonids seem to have lived as tree-dwelling species with highly varied diets.

The location of the center of procyonid radiation in the New World in not known. Jon Baskin, a paleontologist who has performed pivotal analyses of their fossils, postulates that the hub of their early evolution was in "Middle America," a term for tropical Mexico and Central America. His belief is based on the distribution and types of specimens that have been found in different areas. Nevertheless, fossils from this family are exceedingly rare, in part because these animals have never been either diverse or numerous. In addition, during virtually all of their evolution, most of the regions that they inhabited seem to have had lush, subtropical vegetation. Such environments are among the least suitable for fossilization because animal remains readily decompose in their moist conditions.

The earliest members of the raccoon family that ventured into South America presumably evolved from late Miocene procyonines, perhaps from such types as *Arctonasua* or *Stromeriella*. One named *Cyonasua* was probably the first in this lineage to arrive there. Although this name means "doglike coati," these pioneers apparently were not the direct ancestors of the modern coatis. They may have been *waif* immigrants, or those that travel by uncertain means such as by floating on rafts made of vegetation. Indeed, their early remains have been found on coastal Ecuador. However, South American ground sloth fossils have been unearthed in many late Miocene North American sites, suggesting a two-way route between the continents existed at that time. Therefore, these doglike coatis may have also traveled to South America on land. They first appeared there about 7 million years ago and stayed until approximately 2.5 million years ago, into the Pliocene epoch that spanned from about 1.8 to 5 million years ago. As discussed, their ancestors had been shifting toward less reliance on meat. Such feeding versatility was surely a factor in these doglike coatis flourishing in South America for nearly 5 million years.

One of *Cyonasua's* descendants, *Chapmalania*, was approximately the size of a giant panda and may have also subsisted on shoots and roots. None of its direct heirs, however, seem to have survived. The large procyonids might have been outcompeted by the North American bears that traveled to South America following the formation of a Central American crossing in Panama about 2.4 million years ago. Today's South American procyonids descended from the migrants that also wandered across this land bridge. They include members of many present-day forms, such as *Procyon* (the raccoon), *Nasua* (the coati), *Bassaricyon* (the olingo), and *Potos* (the kinkajou). Most were probably woodland or forest creatures, though again, their fossil record is sketchy. The dates of this migration are also unclear. Their dispersal from Central America could have started about 2.4 million years ago and may have continued until fairly recently. In general, the raccoon family's migration into the world's southern continents has been limited. No evidence indicates that any of the Eurasian forms crossed into either Africa or southeastern Asia.

As suggested, the histories of the coatis and raccoons are closely intertwined. Both groups probably descended from a late Miocene animal named *Paranasua*. They appear to have diverged from one another by between 5.2 and 6 million years ago; the oldest fossil evidence for the raccoon genus *Procyon* is from this time. Raccoons were clearly established in the central Great Plains by the middle of the Pliocene, which again lasted from 1.8 to 5 million years ago. By about 1.5 million years ago, several kinds of raccoons ranged across the present United States. This was in the early Pleistocene, the epoch that lasted from about 1.8 million years ago until 11,000 years ago. These raccoons included such species as *Procyon priscus* of Illinois, *Procyon simus* of California, and *Procyon nanus* of Florida, each of which is different from today's raccoons.

Even before the raccoon had appeared, a number of other procyonid-like omnivores existed. In the fossil record, they display an interesting pattern, characterized by the arrival of a genus, followed by its extinction, and then its replacement by another with a corresponding ecological role. This *multiple evolution* of parallel forms, or *ecomorphs*, ultimately culminated in the arrival of the raccoon. Such a pattern of similar types of species succeeding one another suggests that the number of ways in which they can adapt to a particular set of circumstances may be limited. This has also been noted in the repeated occurrence of several kinds of saber-toothed cats. In the case of the raccoonlike animals, perhaps these mid-sized omnivorous Carnivores can succeed only with a few similar body types.

As discussed, two subfamilies, the Ailurinae and the related extinct Simocyoninae, remained in Eurasia after the other procyonids had migrated to North America. The only living ailurine, the red panda, is one of the most controversial of all species in regards to its affinities. Its connection to any present family, or whether it even has one, has been hotly debated for many years. Scientists using evidence from anatomical, paleontological, and molecular genetics studies have all examined this problem. Some assert that the red panda belongs in its own family, the Ailuridae, whereas others believe that it and the giant panda should be placed together in a separate one. Researchers have also suggested that the red panda belongs in the bear family. (Most experts at least agree that the giant panda is a bear.) Nonetheless, it has been persuasively argued that the red panda should be included in the raccoon family. Based upon its dental characteristics and other features, as well as recent convincing molecular analyses, it has been recognized as the sole surviving species of the Old World procyonid subfamily Ailurinae.

It seems that the red panda's oldest identifiable direct ancestor is *Parailurus*, which appears to have lived across northern Europe and western North America in the late Miocene and the early Pliocene, the epoch beginning about 5 million years ago. Given such a broad distribution for its predecessor, it is difficult to know whether the red panda evolved from European mammals that wandered into Asia or possibly from later North American species that immigrated there. Its more recent ancestors have also been discovered in Pleistocene cave sites in China. Now the red panda is restricted to the mountain forests of several Asian nations. Additional information about its distribution and conservation is provided in the following chapter.

Finally, some molecular genetics studies contend that the splitting of the New and Old World procyonids occurred within 10 million years after the raccoon and bear families had diverged. Yet, as pointed out, many authorities agree that the procyonids are actually more closely related to the weasel family than to that of the bears. Other molecular investigations have proposed that this major split within the raccoon family occurred about 28 to 30 million years ago. Such an ancient divergence within this group might at first seem reasonable given their present enormous geographic separation. However, the fossil evidence does not support such an early date for this split because the first definitive members of the family had barely emerged by then. More likely, the division between the New and Old World procyonids occurred closer to the time that the family first appeared in North America, about 16 to 19 million years ago.

Several aspects of the raccoon family's evolution are clearly controversial. The most significant matter of contention involves the very basic issue of determining which family is the closest to theirs and is thus the one with which it shares its most recent common ancestor. The paleontological and anatomical evidence that supports the weasel family Mustelidae in this role seems convincing. Although certain molecular analyses indicate that raccoons and bears have a strong affinity, such techniques may yet reveal that the procyonids have an even closer association with the mustelids.

3

Today's Raccoon Family

The raccoons and their relatives are formally classified as the "Family Procyonidae Gray 1825." The last part of this designation reflects that John Edward Gray first proposed a similar name for this family in 1825; however his term for it was "Procyonina." Another individual, Charles Lucien Bonaparte, apparently first used the term Procyonidae in 1850.

The Procyonidae is a rather small family, consisting of only about 7 genera and 20 species. Its exact numbers vary depending on whether certain forms are regarded as distinct and if the red panda is included. In the Western Hemisphere, procyonids occur from Canada, through Central America, to Argentina. They appear in much of North, Central, and South America. The red panda of Asia, if it truly belongs to this family, would be its sole Old World representative; this panda's limited distribution is described later. The procyonids thrive in a great variety of habitats and climatic conditions. They occur in such different environments as humid tropical rain forests and arid semideserts. Raccoons are especially adaptable; the other family members tend to be considerably more limited in where they live.

One factor, though, is generally common to all of them: they mainly live in areas with suitable tree cover. Trees are vital to procyonids for several reasons. First, each species readily climbs trees to escape from danger. Most also give birth to their young in tree nests, the exceptions being those raccoons that take advantage of alternatives such as ground nests or those that live in relatively treeless locales. Trees also generate a large proportion of the nuts, berries, and fruits that are critical components of their diets. Procyonids are in turn important to their forest communities. As they are primarily *frugivores* (fruit eaters), they probably disperse the seeds of various trees after they eat their fruit.

Because the species in this family are few and somewhat similar, reviewing their common traits is simple. Its present members are all relatively small compared to many of the species in the Order Carnivora. Even the largest procyonid has been described as being barely bigger than a well-fed house cat (which is easily visible when a raccoon is wet). As revealed by their predilection for tree cover, they are largely *arboreal* or *semiarboreal*, meaning that they spend a considerable part of their time in trees. Most tend to be either omnivores or frugivores. They are also usually nocturnal; their activity occurs mostly or completely at night. Their style of locomotion is either *plantigrade* or *semiplantigrade:* the former term means that, like humans, they walk on the soles of their feet with their heels touching the ground; a semiplantigrade animal walks with its heels elevated.

Animal families usually have much greater diversity among their members than the procyonids do, especially in regard to their size or eating habits. Surely if one were to consider some of the raccoon's prehistoric cousins, such as the supposedly digging *Zodiolestes* or the bear-sized *Chapmalania*, then its family would include several other types. In any event, distinguishing the members of this or any family by certain unifying traits, or those that are common to all of them, is crucial. The procyonids are indeed distinguished by specific structural features, although only highly skilled anatomists can detect many of these. As discussed, the sprameatal fossa, a depression in the base of the skull, is considerably large and otherwise unique in this family. In some of its extinct and living members, this structure varies from its characteristic form. However, it might have developed differently in species whose ancestors had initially exhibited the structure's more typical form. As mentioned, it does not develop beyond an early stage in the red panda.

Because of concerns about the defining nature of the sprameatal fossa, its presence alone may be insufficient to identify the family's members. Fortunately, other traits can be used to distinguish them. As mentioned, the carnassial teeth of both procyonids and bears are the least developed for meat eating in the Carnivora. Furthermore, the procyonids' upper molars consist of four rather rounded cusps, a tooth type that corresponds to their diets. Such teeth are referred to as *quadrate,* because they have four cusps, and *bunodont,* given their rounded nature. In addition, their fourth upper premolars (carnassial teeth) also have an extra fourth cusp.

Another distinguishing trait of this family is that its members lack what is known as an *X anastomosis* in their cranial circulation. This term refers to a structural arrangement in which certain blood vessels in the head are joined together at the area where they cross. Because this pattern is present in both the dog and weasel families, the procyonids are actually defined by its absence. Various other definitive skull characteristics have been described, but these descriptions are of such a highly technical nature that one would need a short course in anatomy to decipher them. The skulls of the procyonids encompass a great diversity of types, and often have features whose measurements considerably overlap with those of species in other families. As a result, it is difficult to use some of them to identify members of the raccoon family.

The various procyonid species are described in the following sections. Certain body measurements or distributions are not provided for some because these have not been studied sufficiently to present such information. Finally, conservation issues for individual species and for the family in general are considered.

THE OLINGOS AND THE KINKAJOU

The first group of procyonids considered is the olingos of Central and South America. Their range extends from southern Nicaragua west of the Andes to northern Ecuador. To the east of the Andes, they probably occur from Venezuela to Bolivia, though this has not been confirmed. Olingos are absent from the *llanos* or plains of Colombia and Venezuela but may exist in the western and northern areas of the Amazon basin where the Amazon River's tributaries are most concentrated. Except for a single record from Bolívar, Venezuela, they are apparently absent or rare from the eastern part of northern South America. They also occur inland west of the Panama Canal Zone.

Characterized by rather small ears and an extremely long, lightly banded and somewhat flattened tail, the olingos have a thick, soft fur and sharply curved claws. They are extremely agile, commonly running and jumping through the trees, feeding as they travel. Olingos probably eat fruit, invertebrates, and small vertebrates. Their head and body length ranges from 350 to 475 mm (13.8 to 18.7 inches), and their tail is an additional 400 to 480 mm (15.7 to 18.9 inches). They weigh between 970 to 1,500 g (2.1 to 3.3 pounds). Olingos appear to be restricted to multistrata, tropical evergreen forests that grow from sea level to heights of less than 2,000 m (6,560 feet). (The term *multistrata* refers to a layered effect resulting from vegetation growing at different heights.) Little is known about their ecology or behavior. However, they appear to be primarily arboreal and nocturnal, spending much of the day on nests of dry leaves within hollow trees. They are thought to be highly dependent upon intact tropical humid forests, especially rainforests. They do not seem to be able to adapt to either disturbed or *secondary* forests, those that arise after the original forest is cleared. They also do not appear to prosper in plantations or gardens. Therefore, they are highly vulnerable to deforestation.

The olingos include the following species: *Bassaricyon gabbii*, the bushy-tailed or common olingo; *Bassaricyon alleni*, Allen's olingo, also known simply as the olingo; *Bassaricyon beddardi*, Pocock's olingo, which is also just called the olingo; *Bassaricyon lasius*, Harris's olingo; and finally *Bassaricyon pauli*, the Chiriquí olingo. Researchers disagree considerably about their taxonomic status. Some authorities have concluded that all of them belong to the same species. Another opinion holds that only two species actually exist: *Bassaricyon gabbii* (the bushy-tailed or common olingo), which includes *B. lasius* and *B. pauli*; and *B. alleni* (Allen's olingo), which includes *B. beddardi*.

The bushy-tailed olingo occurs in Nicaragua, Costa Rica, Panama, western Colombia, and western Ecuador, though its distribution is not well understood. It is honey brown and has a banded tail, though the pattern and extent of banding varies. It seems to prefer evergreen forests and primary tropical forests near water. They may be able to live in secondary vegetation or plantations like the kinkajou (which is considered in the following discussion). Yet some observers maintain that they are hardly ever encountered near human development. Deforestation is thus regarded as a major threat to their survival, as exemplified by their apparent demise in northern Nicaragua. In the early 1960s, the forests there seemed to have had significant populations of these olingos. But during the last 40 years, a considerable amount of this region has been cleared for agriculture, grazing, and logging. Warfare in this nation has likely taken a heavy toll on this species as well. Some of the higher montane forests and protected rain forests may still contain bushy-tailed olingo populations.

The habits and ecology of Allen's olingo appear to be similar to those of the bushy-tailed olingo, though not much is known about either species. Allen's olingos range from Ecuador to east of the Andes and from Peru to Bolivia's Cuzco Province. They may also exist in Venezuela and

Brazil, though this is not clear. Similarly, though they are reported to be rare in Bolivia, estimates of their numbers are not reliable. Deforestation is probably also a major threat to their survival. In protected and unprotected areas, Allen's olingo likely faces increasing pressure from the illegal removal of timber, cattle grazing, poaching, and encroaching human settlements.

The very limited information available for the other olingos indicates that their natural histories as well as potential threats to their survival are comparable to those of the aforementioned species. Pocock's olingo, which looks similar to the bushy-tailed olingo, has been reported to be in Venezuela and Brazil. Though there is a record of this species occurring in Guyana, it is probably inaccurate. Deforestation affects most of their supposed range and human activities imperil many of the protected areas in which they might occur. Harris's olingo appears to have a very limited distribution in Costa Rica, near the source of the Rio Estrella in southern Cartago, at an altitude of about 1,500 m (4,920 feet). The last species in this group, the Chiriquí olingo, is only known from a small area between the Rio Chiriquí Viejo and Rio Colorado in Panama's Chiriquí region. In addition to other threats to these olingos, Panama's internal strife may have impacted them; they are not legally protected in this area. As of 2000, the World Conservation Union (hereafter referred to as the IUCN) classified Harris's and Chiriquí olingos as endangered because each has an estimated population of fewer than 250 mature individuals. This organization, previously known as the International Union for the Conservation of Nature, is a prominent conservation organization headquartered in Switzerland.

Because of their similarities and frequent co-occurrence, the kinkajou, *Potos flavus*, is often described in connection with the olingos. All are nocturnal species that live in Central and South American forests. They all have relatively blunt faces with large eyes that may be advantageous for climbing trees at night. Though they are commonly observed feeding together in fruit trees, such as those of figs, kinkajous can usually be distinguished from olingos. The kinkajou is much larger and stockier; although they have been reported to be 50 percent heavier than an olingo, a kinkajou may weigh almost three times as much. They weigh from 1,400 to 4,600 g (3.1 to 10.1 pounds), with males usually being larger. Their head and body length is between 405 and 760 mm (15.9 to 29.9 inches), and their tail is 392 to 570 mm (15.4 to 22.4 inches). As indicated, the olingos have a honey brown color, particularly the bushy-tailed species. The kinkajou's soft wooly fur varies from reddish brown to tawny olive or yellow tawny. Their underparts range from tawny yellow, buff, brownish yellow, to yellowish orange. Though the ranges of their tail lengths are comparable, the kinkajou's is often smaller, shorter haired, and usually tapers to a black tip. Moreover, it is prehensile; it can be used for grasping. Alternately, the olingo's tail is relatively long for its body size, bushy, lightly banded, and is not prehensile.

Kinkajous are found throughout the Neotropics, from Mexico to Bolivia. They occur from Mexico to the east and south of the Sierra Madre; along the central and southern Mexican coasts; southward through Beni, Bolivia, which is east of the Andes; and deep into the Mato Grosso Plateau in Brazil. Although it has such a wide distribution, its inland occurrence is limited in much of Central America, and it has generally been rare in other parts of its range as well.

Despite its similarities to the olingos, the kinkajou is so anatomically distinctive that it is placed in its own genus. Although other mammals possess prehensile tails, it is a decidedly odd feature for a Carnivore. The binturong of the Family Viverridae is the only other one with such a tail. Clearly, the kinkajou's extreme arboreal nature is facilitated by having this type of tail, as it can hang from it to reach otherwise inaccessible fruit. In some areas, this behavior seems to have resulted in their being mistaken for monkeys. Kinkajous also differ from their family members in that they have one less premolar on each side of their mouth and they lack an anal scent gland; they have a scent gland on their chest and belly region instead.

As mentioned previously, the kinkajou is so unique that it is considered to be the only member of the Subfamily Potosinae. Some, however, argue that the olingos are similar enough to be placed in this subfamily. In also either case, the kinkajou and the olingos are believed to be direct descendants of the family's most ancient subdivision, and as such, have been considered to be the most primitive procyonids.

The term primitive, however, can be misleading; in this case, it refers only to these species being the most connected with the family's oldest branch. In fact, the olingos and the kinkajou have departed from a predominantly carnivorous feeding mode to a greater extent than their relatives, with the possible exception of the red panda. Their flat carnassials serve largely to crush fruit rather than to slice meat. The olingos augment their fruit and nectar diet with small mammals, birds, and insects, often foraging alone or in pairs. The kinkajou lives almost entirely on fruit, especially figs, though it may occasionally eat insects. It also uses its tongue to probe floral nectaries to reach their sweet liquid. Kinkajous may be important in the dispersal of seeds, especially those of certain fruit trees. They may feed in large groups when food is plentiful, but they have also been described as being mostly solitary. Primarily nocturnal and almost entirely arboreal, they have been observed to spend the day inside hollow trees.

Several of the olingos in a Costa Rican study appeared to be *monogamous*. This is the arrangement in which a male and a female form a pair bond and stay together at least through the mating season. An olingo's pregnancy lasts for about 73 to 74 days and she will likely have just one offspring per litter. The kinkajou remains pregnant for quite a while longer, from 112 to 120 days, and the female also usually gives birth to a single young.

Given their strong reliance on trees, kinkajous require a closed-canopy forest, one with little space between the treetops, such as those in Central America and the Amazon River basin. Hence, they do not occur further to the south and east of their range, where the land is more open and more arid. It is doubtful that they exist on Venezuela's savanna, though this has been noted on some maps. No evidence indicates that the kinkajou is threatened. Nevertheless, given their dependence on closed-canopy forests, their numbers are bound to decline with increased disturbances, especially deforestation. They may also be targeted by the pet trade. Though some indigenous tribes hunt kinkajous, to use their hides for drums, for example, this does not appear to be a significant problem.

THE RINGTAIL AND THE CACOMISTLE

The genus *Bassariscus* includes two species: the ringtail or ring-tailed cat, *Bassariscus astutus*, and the cacomistle, *Bassariscus sumichrasti*. Though the former is also referred to as a cacomistle, it is more commonly known as the ringtail. Its scientific name is a Greek-Latin mixture that means "clever little fox." The word cacomistle stems from a Mexican Nathuatl Indian term, *tlacomiztli*, which means "half mountain lion." Because early North American miners kept them in coal and gold mines to control rodents, ringtails became known as the "miner's cat." They were reputed to be better mousers than cats. The ringtail is the more northern of the two species, ranging from southwestern Oregon through Colorado, and south to the Guerrero, Oaxaca, and Veracruz provinces in central Mexico, where they overlap with the cacomistle. In the twentieth century, ringtails have apparently occurred in such distant states as Kansas, Arkansas, and Louisiana, and have even been reported in Ohio and Alabama. Oddly, some of these occurrences may have resulted from individuals that boarded trains. The cacomistle is found east of the Sierra Madre in central Mexico and south of this range into southern Mexico.

The ringtail or ring-tailed cat, *Bassariscus astutus.* Drawing by Elizabeth Dewitte.

They extend further southward to extreme western Panama and inland south of Guatemala.

The cacomistle and the ringtail are the only two members of the family that have retained carnassial teeth that are both sharp and *sectorial*, or capable of slicing. Indeed, among the modern procyonids, the cacomistle has been depicted as the species that most closely resembles the family's earliest members, both in its climbing skills and largely carnivorous diet. This characterization might appear to conflict with the earlier presented speculation that members of the Subfamily Procyoninae, to which the cacomistle belongs, have departed the most from the family's ancestors. Recall, however, that this conjecture referred to other members of this subfamily: the coatis and the raccoons. Such a distinction between these groups may support placing the cacomistle and ringtail in their own subfamily, the Bassariscinae, after all. Furthermore, the notion that the cacomistle most closely resembles the family's earliest members does not necessarily negate the assumption that the kinkajou and olingos are the more direct heirs of these ancient procyonids. Perhaps the cacomistle has only more recently evolved to resemble the family's original types, whereas the kinkajou and olingos are actually more directly related to them. Much uncertainty remains about these classifications.

Ringtails and cacomistles are among the smallest procyonids. Both are highly attractive, having foxlike faces and rather large ears. Those of the ringtail are rounded, whereas the cacomistle's are tapered or pointed. The ringtail's head and body length vary from 305 to 420 mm (12 to 16.5 inches); its tail is an additional 310 to 441 mm (12.2 to 17.4 inches). It weighs from 824 to 1,338 g (1.8 to 3.0 pounds). The cacomistle's head and body length ranges from 380 to 470 mm (15.0 to 18.5 inches), and its tail is 390 to 530 mm (15.4 to 20.9 inches) long. It weighs approximately 900 g (2.0 pounds); it has not been studied sufficiently to know its weight range. Both species are recognizable by their long, bushy tails, which are strongly banded with black-and-white or tawny rings. The ringtail's has seven to nine rings. For similar-sized individuals, the cacomistle's tail is generally the longer of the two.

The ringtail's upper body parts are a buff color, intermingled with black or dark brown hairs. Its underparts are white or white highlighted with buff. Its eyes are circled with a black or dark brown color that is set within larger white rings, creating a masked appearance. The head also has white to pinkish buff-colored patches. The cacomistle's upper parts and sides are tawny brown or buffy gray to brown. It is yellowish white underneath. Their eyes are also outlined by black rings set within broader paler circles. The cacomistle's feet have naked soles whereas the ringtail's are hairy.

Both species have long legs and are very agile. Extremely nimble climbers, they use their long tails for balance. They each have the extraordinary ability of being able to rotate their hind feet at least 180 degrees. Together with footpads that provide good traction, they can rapidly scurry down precipitous cliffs and tree trunks. Because of its dexterity, the ringtail is sometimes referred to as a *mico de noche*, or "night monkey," in Mexico. It is the only procyonid with semiretractile claws. The ringtail has been known to squeeze into crevices by pressing all of its feet against one wall and its back against the other. They can glance off of smooth surfaces to gain momentum in their movements. Both species are skilled hunters and may kill their prey by pouncing on it and biting the neck. Perhaps this is why a name that translates as "half mountain lion" was bestowed upon the cacomistle.

Ringtails are solitary and highly nocturnal. Their diverse habitats include rocky or cliff areas, scrub, desert, dense riparian zones (those along rivers or streams), and evergreen forests. They may also be common in some metropolitan areas. Ringtails do not occur above about 2,800 m (9,187 feet). They den in rock crevices, hollow trees, cabins, and Native American ruins. The cacomistle has not been studied as much as the ringtail. Although it is also assumed to be nocturnal and solitary, it may be more arboreal and live in wetter forests than the ringtail. Their preferred habitats seem to be lowland and montane rainforests, though they have been documented in wet evergreen forests, seasonally dry forests, scrub, and secondary forests. The cacomistle appears to mostly use the middle and upper levels of tropical forests. It may only den in trees and evidently rarely ventures onto the ground. Both species feed on rodents, birds, reptiles, and fruit. Large insects such as grasshoppers often constitute their primary food items. At least in the case of the ringtail, much of its food is succulent or juicy. The moisture from such items, coupled with the ringtail's ability to produce highly concentrated urine and therefore retain water, enables it to live in dry environments.

Ringtails generally hunt alone. They have been reported to live in pairs within about 100 ha (about 250 acre) territories, though the size of these areas likely varies across their range. Their gestation period lasts for 51 to 54 days; it is the shortest pregnancy within this family. They bear from one to four young but usually have two or three per litter. Although the fathers may bring food to their litters, the amount of paternal care is not known. Cacomistles probably also forage alone. They also seem to lead solitary lives, particularly outside of the long breeding season. Surprisingly, their gestation period is much longer than that of the ringtail, lasting from 63 to 66 days. They also differ in that the cacomistle usually produces only one young per litter.

Unfortunately, the cacomistle is yet another species that is threatened by deforestation, other types of habitat disturbance, and the resulting fragmentation of its populations. Though its status is unknown in many areas, it may have never been common. Apart from some areas in Mexico, such as the remaining forests in Veracruz and Honduras, it might be seriously threatened over much of its range. It may be distributed only in patches throughout much of Mexico and Guatemala. They are believed to be extinct in west Panama and rare throughout that country. They also seem to have disappeared from much of the Costa Rican plateau, and no information is available on their status in Nicaragua. In Honduras and Mexico, cacomistles are hunted for their fur and because they are believed to kill chickens. However, captive individuals are said to be afraid of birds larger than doves. Some indigenous people such as the Lacandones, descendants of the Maya in southern Mexico, Guatemala, and Belize, hunt them for food.

The ringtail's primary hazards are automobiles and trappers, though they are also preyed on by hawks, eagles, and owls. They are legally trapped for their fur in Arizona, New Mexico, Colorado, and Texas, and many are caught in traps set for other furbearers, such as foxes and raccoons. In Texas, about 45,000 to 50,000 were trapped annually from 1975 to 1985. The total number trapped has declined following a 1979 peak, when about 135,000 were sold. In recent years, approximately 4,000 ringtails have been taken annually in Arizona, whereas about 1,000 were taken in New Mexico. In the 1991–1992 season, only 5,638 were harvested in the entire United States. Commercially known as "California mink" or "civet cat," their coats are typically thin and often fade; thus trapping them seems unjustified. Their pelts have usually sold for less than five dollars, though they have brought in as much as twelve dollars. Because their population biology is poorly understood, it is difficult to devise appropriate harvest regulations for them.

THE COATIS

The coatis include a handful of species, all of which have extended, piglike snouts and long, striped tails. The genus name for most of them, *Nasua*, is derived from the Latin *nasus*, meaning nose. As was the case for the olingos, taxonomists disagree about the status of some of them. *Nasua nasua*, alternately known as the South American, ring-tailed, or brown-nosed coati is clearly distinct from *Nasua narica*, the Central American or white-nosed coati. Another, *Nasua nelsoni*, the Cozumel Island coati, is only found on this island off of Mexico's Yucatán Peninsula. Questions remain about whether it is indeed a genuine species. Last, *Nasuella olivacea* is the lesser, little, or mountain coati. Some authorities have doubted the validity of placing this species in a separate genus.

The name *coati* is of Tupian Indian origin. Based on the words *cua* for belt and *tim* meaning nose, together they produce a term reflecting the animal's habit of sleeping with its nose tucked in its belly. This word has also been thought to be derived from the Guarani Indian term *Kuat-l*. However, these groups may have had similar terms for the same animal. The word *coatimundi*, now used to refer to all coatis, originally described a solitary male coati. It stems from the Brazilian *coati monde*. The coati is thought to have been a fertility symbol for the Maya. Presumably, it was kept both as a pet and a food source and may have been eaten only by women. Other indications of its significance for different cultures are revealed in its various local names.

The coatis of the genus *Nasua* are found in many types of forest habitats, from tropical rainforests to dry scrub communities. Though highly adaptable, at least in the case of the white-nosed coati, they are rarely observed in open grasslands or deserts. The range of this genus extends from far southern Arizona, New Mexico, and Texas, south through Central and

The white-nosed or Central American coati, *Nasua narica*. Drawing by Elizabeth Dewitte.

South America, to Argentina and Uruguay. They appear to be absent from the llanos of Venezuela. The white-nosed coati ranges from Arizona and parts of southern New Mexico, south through Mexico (except the Baja Peninsula) and all of Central America to Panama, to the west coasts of Colombia, Ecuador, and Peru west of the Andes. Although solitary males have been reported from southwest Texas, they probably do not breed there. The South American coati occurs east of the Andes from Colombia and south to Uruguay and Argentina.

Though the white-nosed and South American coatis can be distinguished by several skull characteristics, these are not readily observed. For example, their most notable skull difference is that the white-nosed coati's palate is depressed along its midline whereas the South American's is flat. Both are usually reddish brown to black above and yellowish to dark brown underneath, and have black and gray facial markings. On each, the muzzle, chin, and throat are generally whitish and their long, pointed muzzle has a rather mobile tip. The tapering tail is banded and longer than the head and body. The feet are blackish.

The white-nosed coati is typically pale brown to reddish above, and its eyes are bordered by a reddish to brown mask. Yellow or white hairs occasionally distinguish the adult male's shoulders. Though fairly adaptable to various wooded habitats, they do require some cover, most likely for protection from predators. Its head and body measures from 430 to 700 mm (16.9 to 27.6 inches), and it has a striped tail that is an additional 420 to 680 mm (16.5 to 26.7 inches). The weight range for the genus is 3,500 to 6,000 g (7.7 to 13.2 pounds); males are usually larger than females.

Whereas the South American coati's color is highly variable, from reddish orange to nearly black with a gray face, they are usually a dark reddish brown, with an off-white chin and throat that blends into a pale yellow underside. They have a long, tapering blackish brown tail with yellow rings. Its length measurements roughly coincide with those of the white-nosed coati. Its basic ecology and social group dynamics are also thought to be quite similar to those of its northern cousin.

The third species in this genus, the Cozumel Island coati, only occurs on this island. They are alleged to be the descendants of white-nosed coatis possibly introduced by the Maya, as Cozumel was an important location for worshiping their fertility goddess, Ix Chel. As mentioned, the coati may have been a Mayan symbol of fertility. Because of their close connection, some authorities consider the Cozumel Island coati to be a subspecies of the mainland form, the white-nosed coati. Nonetheless, the island form is generally smaller and has a silkier fur. Hardly anything is known about its habitat requirements, though they are likely similar to those of the white-nosed coati.

The mountain coati has the most restricted range of any procyonid genus. It only occurs in high-altitude montane forests, at elevations greater than 2,000 m (6,560 feet). It is found in the Andes in Colombia, western Venezuela, Ecuador, and possibly extreme northern Peru. It is usually grayish brown with darker legs and feet, and has a banded dark brown and yellowish gray tail. Mountain coatis differ from the other coatis in several ways. They are smaller and thinner, which is often the case when comparing animals adapted to high altitudes with their close relatives. Only about half the size of the white-nosed coati, a mountain coati's combined head and body length is about 360 to 394 mm (14.2 to 15.5 inches), with a tail that is an additional 200 to 242 mm (7.9 to 9.5 inches). Its tail is also proportionately shorter than those of the other coatis. Furthermore, its dentition appears to be more adapted to insectivory. Very little is known about its ecology or behavior. Much of its habitat is threatened by a rapidly expanding human population, conversion of the *cloud forest* (tropical, mountainous forests characterized by heavy rainfall and persistent condensation) to agricultural lands, and the planting of pine trees in an area called the *paramo* (defined as "wasteland"). Even a small degree of habitat change could be very detrimental to this species given its restricted range.

For a group of animals that appears to be ungainly, coatis are actually quite agile and are good climbers. At least some species have reversible ankles that supply additional traction as they descend headfirst from trees. They have strong forelimbs and long claws that assist them both in climbing and searching for prey. The long, slender tail has a counterbalancing function while climbing. When alarmed, coatis may seek refuge in trees. They also sleep in the treetops, possibly for safety. Essentially diurnal, they may be largely terrestrial, arboreal, or a little of both depending on the habitat.

Though they have highly variable group sizes, coatis commonly display a distinctive pattern of social behavior. Adult females live in groups of up to 30 individuals, though they and their young commonly form permanent groups of up to 12. For much of the year, the adult male is solitary. It only joins the family group during the breeding season, when the females become less aggressive toward intruders. When a male is accepted into this group, he becomes fully subordinate to the females. Normally, he is ousted from the group shortly after mating, though some may continue to associate with the females. Their breeding season coincides with the period of maximum fruit abundance. The reduced competition for food that could occur at this time may enable a male to remain with the group. Gestation in coatis lasts for 70 to 77 days; their litter sizes range from two to seven poorly developed young.

Coatis have the unusual habit of holding their tails aloft while foraging, which may help group members to stay within sight of one another in dense vegetation. They eat fruit, invertebrates such as beetles and worms, and small mammals and other vertebrates, including frogs and lizards. Adult males have been reported to prey on large rodents and may eat young coatis outside of the breeding season. Coatis search for fruit high in the forest canopy or for animals along the forest floor by poking their sensitive snouts into crevices, knocking over rocks, and ripping apart dead logs.

In New Mexico, the white-nosed coati is classified as an endangered species and thus receives full protection. Yet in Arizona, where the only substantial populations in the United States occur, they are subject to year-round hunting. It is curious for two neighboring states to have such contrasting policies toward the same animal. Elsewhere in their range, coatis do not appear to receive any official protection. Their U.S. populations are gradually becoming isolated from those to the south, which is at least in part due to the apparent severe decline of many Mexican populations. This trend could lead to their demise in nearby areas in the United States.

Substantial development has taken place on Cozumel Island as a result of its transformation into a popular holiday resort. Given such habitat destruction and their limited distribution, its coatis face considerable threats. This activity could threaten other species that depend on the island's restricted and fragile forests, such as the Cozumel Island raccoon. Another potential threat to the Cozumel Island coati is the introduction of pet white-nosed coatis to this island. Should these escape or be abandoned, they could contaminate the island coati's gene pool. Even if they actually belong to the same species, by now the Cozumel Island coati is quite distinct. If they were to breed with released pets, it would be difficult for them to persist as a unique type of coati. The IUCN classifies it as an endangered species because fewer than 250 mature individuals exist.

In some areas, such as in Argentina's Parque Nacional Iguazu, coatis may be the most abundant Carnivore. Yet their populations are often at risk as their habitats have become threatened by deforestation for agriculture and logging. Coatis are also susceptible to canine distemper and rabies. In addition, they are hunted throughout their range for fur and meat. In Peru, their parts are eaten for their supposed aphrodisiacal powers. In various places, they are caught in traps set for other furbearers. In Arizona, the law requires such coatis to be released, but they are frequently killed anyway. They may also be victims of traps set for species in predator control programs. Given their small and unstable populations in the United States, efforts should be made to discourage such killings through trapper education.

THE RACCOONS

The last major group examined is the raccoons. With the greatest range of all of the groups in their family, they extend from Canada to Argentina.

Seven different types of raccoons are recognized, though again authorities differ on whether many of them, particularly the various island forms, are distinct species. All of the raccoons are in the same genus, *Procyon*. The raccoon has two *subgenera*, or small groups of closely related species. The subject of this book, the raccoon or common raccoon, *Procyon lotor*, and several island forms, mostly from the Caribbean, belong to the subgenus *Procyon*. The other subgenus, *Euprocyon*, includes only the crab-eating raccoon, *Procyon cancrivorus*. Of the seven species, only the common and crab-eating raccoons are widely distributed; the others are restricted to particular islands. The common raccoon is widespread through much of North and Central America. As this book's main subject, it is characterized in detail throughout these chapters. Many of the raccoons are similar in size, and unless stated otherwise, have dimensions that are consistent with those presented for the common raccoon in the next chapter. The southern forms of the common raccoon, however, are typically smaller, a characterization that often also applies to the island species. Finally, the common raccoon has numerous subspecies; a detailed description of each is provided in Chapter 5.

Each of the island raccoons has a very restricted range. The Barbados raccoon, *Procyon gloveralleni*, only occurs on Barbados, an island in the Lesser Antilles group. Small and dark, it is heavily overlaid with black and its underparts are thinly covered with a buffy grayish hue. Atop the head, it is a buffy gray mixed with black. Yet this raccoon's upper pelage is usually a light ochraceous buff, which is most evident on the nape of its neck and shoulders. The mask is continuous across its face. Its skull is rather short and seems to be somewhat delicate. Although it was probably once abundant, especially on the island's rugged south side, the Barbados raccoon was last sighted in 1964 and is extinct according to the IUCN.

The Guadeloupe or Guadeloupean raccoon, *Procyon minor*, is another possible raccoon species from the Lesser Antilles islands. It only occurs on Guadeloupe Island, which France administers. Also small and rather dark, it is generally grayish with an ochraceous buff color on the neck and shoulders. It is heavily overlaid with black on its back, producing a grizzled effect. Its underparts are so thinly overlaid with a grayish hue that its light brown underfur shows through.

Although they are widely separated geographically, the Guadeloupe raccoon appears to be most closely related to the Bahama raccoon, the next one discussed. They differ in certain details but their delicate skulls are similar. A recent genetic analysis, however, has indicated that the Guadeloupe raccoon is actually not even a unique species and that it is a variant of the common raccoon. As indicated in the following discussion,

The raccoon, *Procyon lotor*. Drawing by Elizabeth Dewitte.

this may also be the case for the Bahama raccoon. Land use changes on Guadeloupe have resulted in a considerable loss of its native habitat. Now, this raccoon is confined to the island's mangroves and remaining stands of rainforest. (*Mangroves* are tropical maritime trees that often form dense masses along coastal areas. They are characterized by large exposed roots.) Even though it is believed to be rare, the Guadeloupe raccoon is still hunted for food. Yet another threat to its survival is the reported introduction of the crab-eating raccoon (described later), which has also been released on Trinidad and Tobago. As of 2000, the IUCN has classified the Guadeloupe raccoon as endangered because its sole population of fewer than 2,500 mature individuals has continued to decline. It evidently still occurs in the Parc National de Guadeloupe; it has been selected as the emblematic species of this reserve.

The Bahama or Bahamas raccoon, *Procyon maynardi*, is only known to occur on New Providence Island in the Bahamas. It is a small, medium-gray–colored species, somewhat paler than the Guadeloupe raccoon. The gray becomes ochraceous buffy at the nape of its neck and shoulders. It is moderately overlaid with black fur that thins out along the sides. Its underparts are sparsely covered with grayish fur. The top of its head is a grizzled mix of gray and black, and its black mask is interrupted between the eyes. Its skull and dentition are both slender and delicate.

Because of the similarities of their skulls, the Bahama raccoon seems to be closely related to the Guadeloupe raccoon as well as to the raccoons of the Florida Keys. Though its occurrence in the Bahamas may have resulted from an introduction of the mainland form, there are no records of this. Still, doubts have been raised for many years about its being a separate species. It has also been classified as *Procyon lotor maynardi*, which would make it a subspecies of the common raccoon. Further study may result in such a change in status for some of the other island raccoons as well. Development in the Bahamas has undoubtedly had an impact on this animal. It is not legally protected nor are there protected areas that are known to contain it. The IUCN has classified this raccoon as endangered because its severely fragmented population of fewer than 2,500 individuals continues to decline.

The Trés Marías Islands raccoon, *Procyon insularis*, occurs off of the west coast of Nayarit, Mexico, on two of the Trés Marías Islands (Las Islas Marías): María Madre Island and María Magdalena Island. It is a large, pale species with a short, coarse, bristly pelage. Its upper parts are normally a light creamy buff rather than the iron grayish hue of the nearby mainland types of the common raccoon: the Mexican raccoon and the Mexican Plateau raccoon. It has an imposing skull that is considerably more angular than that of the common raccoon. On its back, it is thinly overlaid with black. Its underparts are lightly covered with a pale creamy buff color, permitting its light brown underfur to show.

This raccoon is not common and likely never has been. Its habitat preferences are probably similar to those of the common raccoon. The major threats to its survival are its limited distribution and the islands' inhabitants who both hunt them and capture them for pets. The Trés Marías Islands raccoon is not protected nor does it have protected areas that could enhance its survival. Unfortunately, as of the year 2000, the IUCN also classified this raccoon as endangered because its population contains less than 250 mature individuals. The subspecies on María Magdalena Island, *Procyon insularis vicinus*, is considered to be extinct.

The next raccoon considered is the Cozumel Island raccoon, *Procyon pygmaeus*, which occurs only on this island located off of Mexico's Yucatán Peninsula. As its scientific name implies, it is the smallest raccoon species, often weighing just 3 to 4 kg (6.6 to 8.8 pounds). It has a short, bristly, light buffy grayish pelage that is thinly overlaid with black. The underparts are finely overlaid with light buffy hairs, allowing the light brownish underfur to show through. On its head, it is a grizzled gray and black. Its black facial mask tends to become brownish, mixed with gray, along the middle. It has a relatively flat skull, similar to that of the Campeche raccoon, the common raccoon subspecies on the adjacent mainland. Its fur is also similar, both in color and texture, to that of this subspecies. The Cozumel Island raccoon, however, is noticeably smaller, has a shorter and narrower snout, and strikingly small teeth, differences that suggest it has been separated from its mainland cousin for a long time. It lives in mangrove swamps, but virtually nothing else is known about its natural history.

As mentioned, this island has been substantially developed in recent years, especially around its coastline and close to its mangroves. As is the case for several of the other raccoons, it is not legally protected and the island has no protected areas. As of 2000, the IUCN also classified this species as endangered because of the continued decline of its severely fragmented population of less than 2,500 individuals.

Some authorities believe that all of the island raccoons actually belong to the same species. Yet others consider only the Trés Marías Islands and the Cozumel Island raccoons to be distinct. In any event, all are assumed to be relatively recent arrivals to their islands, though little is known about how they reached them. Humans could have transported their ancestors to these locales, but given their affinity for water they might have embarked on their journeys by climbing aboard rafts of earth and vegetation. Given the small sizes of their ranges, they have probably never been numerous. In each case, island tourism and the resulting dwindling of their habitats has negatively impacted them, as exemplified by the likely extinction of the Barbados raccoon and the endangered statuses of the others.

The crab-eating raccoon ranges from southern Costa Rica and the Isthmus region of western Panama to northern Argentina. It occurs along the east border of the Andes, on Trinidad, and possibly on several other Caribbean islands. It is probably on the island of Guadeloupe where it may have replaced the indigenous Guadeloupe raccoon, one of the island forms just discussed. It is also possible that individuals of the latter species have been misidentified, leading to the belief that crab-eating raccoons occur there.

The color of the crab-eating raccoon varies somewhat. Its back ranges from ashy gray to ochraceous buff to yellowish ochraceous, all of which are usually overlaid with black. (*Ochraceous* is a term often used to describe fur color. It refers to an earthy hue, usually yellow or reddish.) This raccoon's ears, the bands above its eyes (known as *supraorbital streaks*), and the sides of its muzzle are whitish. It has the characteristic black mask, with a middle line extending from the forehead to the nose. Its throat is a grayish white. The underparts vary from pale gray to yellowish or ochraceous. The outer surfaces of its forearms and thighs are usually blackish rather than grayish, as in the common raccoon. Its feet vary from gray to brown. Like its cousin to the north, the crab-eating raccoon has a ringed tail that has seven to eight black or gray and yellowish rings that culminate in a black tip.

Compared to the common raccoon, the crab-eating raccoon has a shorter pelage and lacks a finely textured underfur. Its short, thin, bristly coat causes it to appear smaller than similar-sized common raccoons and makes its fur less valuable. Oddly, the hair on the nape of its neck is directed forward rather than backward as in the common raccoon. In tropical regions, its coat may not undergo a seasonal shedding, or molt. Its claws are also different from those of its northern cousin. The claws of the common raccoon and its closest relatives are both sharper and narrower. They appear to be better suited for climbing and are likely a factor in the common raccoon being more arboreal than its cousin. Lastly, the crab-eating raccoon's teeth are far better suited for crushing hard substances such as crab and mollusk shells. With the exception of their first premolars, their cheek teeth are more massive and have broader, more rounded cusps than those of the common raccoon.

Crab-eating raccoons are *sympatric*, or coexistent, with common rac-

The crab-eating raccoon, *Procyon cancrivorus*. Drawing by Elizabeth Dewitte.

coons in Costa Rica and immediately east of its border with Panama. In these areas, the common raccoons are typically found in mangrove swamps, while its crab-eating cousins occur along the inland rivers. Although crab-eating raccoons seem to be largely restricted to these coastal and riverbank habitats, they are also found in evergreen forests and on the plains. They are only infrequently found in rainforests. They tend to be nocturnal, terrestrial, and solitary. In addition to crabs and mollusks, they eat fish, insects, and amphibians. Little else is known about their ecology or behavior. In general, it seems to be less adaptable to human activity then the common raccoon. It is probably stable throughout South America in suitable habitats though naturally rare in other parts of its range. Threats to its survival include local overhunting, collection of their young for the pet trade, and random shooting, such as for target practice. As is the case for far too many species, they are also threatened by the continued eradication of their forest habitat.

THE RED PANDA

As stated in the previous chapter, the red or lesser panda, *Ailurus fulgens*, is believed to be sole Old World member of the raccoon family. The great French zoologist, Frédéric Cuvier, the brother of Georges Cuvier mentioned in Chapter 1, introduced the red panda to Western naturalists in 1825. He called it "quite the most handsome mammal in existence." Throughout much of the nineteenth century, when Western zoologists spoke of a panda, they were referring to this animal. The word is derived from a French attempt to spell a Nepalese name for the red panda, not the giant one. When the larger panda began to attract more attention, the reddish one became the lesser panda. It is only recently that it has started to shed this diminutive label in favor of red panda.

There are two recognized subspecies of the red panda: *Ailurus fulgens fulgens*, the standard form, and *Ailurus fulgens styani*, the Chinese or Styan's red panda. Though the latter has been described as being larger and darker, each form varies considerably in color and size. Their upper body color varies from a rusty to deep chestnut; it is darkest along the middle of its back. Underneath, they are dark reddish brown to black. Their coat is long and soft, and the bushy tail is faintly ringed. Red pandas have small, dark eye patches; the snout, lips, and cheeks are white. Their head is fairly rounded, and their large, pointed ears are edged with white. Their head and body length ranges from 510 to 635 mm (20 to 25 inches), and the tail is 280 to 485 mm (11.0 to 19.1 inches).

Confined to the Himalayas, the red panda's range stretches from Nepal in the west to Tibet and Szechwan in the south of the People's Republic of China to the east. Outside of China, its range extends to northern India including its northeastern state of Sikkim, Nepal, Bhutan, and northern Myanmar (formerly Burma). They are limited to an altitudinal zone of between 1,500 or perhaps 2,000 m to 4,000 m (4,920 or 6,560 feet to 13,120 feet). In China, they overlap with the giant panda and utilize similar habitats. The red panda is a habitat specialist as it apparently only occurs in mixed spruce-fir forests that have a dense understory of various

The red or lesser panda, *Ailurus fulgens*. Drawing by Elizabeth Dewitte.

bamboo species. They can be uncommon, even in their more recognized refuges such as Nepal's Langtang National Park.

The red panda leads a mostly solitary life. Adult males monitor the limits of their mutually exclusive territories, each of which may overlap those of several females. Consequently in Nepal, males have larger *home ranges* (1.7 to 9.6 km^2) than females (1.0 to 1.5 km^2). A home range is the area in which an individual engages in its various activities; it can vary because of several factors, as are described for the raccoon's home range in

Chapter 9, "Social Organization." Little is known about the red panda's reproduction. Its litter size appears to range from one to four, though they usually bear two young. Like the giant panda, its diet is largely limited to bamboo, occasionally supplemented by grasses, fruit, berries, roots, acorns, lichens, birds, and small rodents. They only feed on the fresh young leaves at the base of the bamboo's stem. Ironically, given their reputation as extreme food specialists, the giant panda may be less particular about eating bamboo parts as it uses both the leaves and the stems. The red panda's flat, broad cheek teeth, with their complex pattern of ridges, resemble those of a herbivore. When not feeding, it spends most of its time in the trees; they are highly arboreal.

Human cultures may have recognized red pandas long before other procyonids. The earliest known depiction of one is from a thirteenth-century Chinese drawing of a hunting scene. In certain Chinese cultures today, the groom may wear a red panda skin at his wedding ceremony. It also has prominence in India. One of its prime ministers, Indira Gandhi, kept them as pets when she was a child. Several years ago, it was the mascot of an international tea festival in Darjeeling, India. The red panda has also been embraced as Sikkim's "national" animal, though this region is actually an autonomous state in northern India rather than a full nation.

The numbers and status of the red panda are unknown throughout much of their historical range, such as in Myanmar, Bhutan, Sikkim, and the Himalayan portion of India. Surveys indicate that they are generally far from abundant and have an uneven distribution. One estimate states that only 300 remain in Nepal, though this figure may be biased due to the study area's low altitude. They could be relatively common in eastern Nepal at altitudes of 2,500 to 4,000 m (8,200 to 13,120 feet) where it is damper and hence better suited for bamboo. Preliminary fieldwork in Wolong National Park and other protected areas in China indicates that they are also rare in this country.

The red panda is primarily threatened by deforestation. The destruction of their forest habitat is not only caused by the usual factors, such as expansion of agricultural land, overgrazing, and use of wood by residents, but also by ecotourism. This may seem paradoxical, as one might assume that nature-oriented travelers and tour companies would be sensitive to potential environmental degradation. Yet a Himalayan hiker can use more firewood in one week than a local person does in a year. The loss of forests has a particularly insidious effect on red pandas, since it readily leads to the disappearance of bamboo, their main food source. To better understand how the red panda reacts to different wood-harvesting practices, studies should be conducted on their bamboo-forest ecosystem.

Another problem they face is that habitat fragmentation and the ensuing segregation of their populations can lead to individuals within a species becoming genetically isolated from one another. Even within Nepal's Langtang National Park, four separate red panda populations exist. Such separation can create a variety of problems for small populations, such as *inbreeding depression* in which reproductive rates decline and susceptibility to disease increases. Moreover, because they are such extreme food specialists, if individuals cannot travel to new areas when their food is depleted, such as when the bamboo dies off after flowering, they could starve. To facilitate interaction between isolated red panda groups, migration corridors have been established between forests. In addition, some cleared areas have been replanted with native vegetation. The red panda is also hunted for its fur, which is primarily sold in local markets. It is likely that many are caught in snares set for other animals, especially the musk deer, which is sought for musk, a key ingredient in perfumes and Chinese medicines.

Conservation biologists are well aware of the red panda's plight. In 2000, the IUCN classified it as endangered because of the ongoing decline of its severely fragmented populations, which consisted of fewer than 2,500 individuals. As of 2001, it also appeared in Appendix I of an influential pact, the Convention on International Trade in Endangered Species of Wild Fauna and Flora (CITES). This appendix lists the most endangered plants and animals. Commercial international trade of such species is normally prohibited among the pact's many signatory nations. The red panda has protected status in Nepal and China, two of the principal countries in its distribution. Because they live in similar habitats, red pandas have probably benefited from the reserve system established for giant pandas. However, the people who come into protected areas in Nepal and China to obtain or use various resources can threaten their populations. Its legal status in several countries is not clear.

Fortunately, hundreds of red pandas are in zoos throughout the world where they are the subject of an international captive breeding program. Given the recent success of this effort, biologists anticipate having viable populations of red pandas within the next few years. Nevertheless, reintroduction of captive animals into their natural habitat can be both expensive and risky. The most critical aspect of the red panda's conservation is for it and its mountain habitats to be better protected and managed.

THE RACCOON IMITATOR

One last animal worth considering is the raccoon dog or tanuki, *Nyctereutes procyonoides*. Despite looking very much like a raccoon, with its masked face and yellowish brown fur, and though the term *procyon* is embedded within its scientific name, the tanuki is actually a canid. Not only does it resemble the raccoon, it even exhibits similar behaviors. They have a predilection for acorns and garbage; undergo seasonal shifts in their diet of small animals, berries, and fruit; and tend to stay in dens in the colder parts of their range, which includes Japan, other areas in Asia, and Europe. It appears that this is a case of convergent evolution. As described in the previous chapter, this occurs when unrelated organisms that have adapted to similar ecological niches evolve to resemble one another. Yet just why the raccoon or tanuki look and act like they do are open questions. Several possible roles for the raccoon's mask and its coloration are offered in the next chapter.

GENERAL THREATS TO PROCYONIDS

The various threats to individual procyonid species have already been addressed. Recently, the common risks to their survival were reviewed in two IUCN reports by Angela Glatston and David Stone. These publications focused on the recommendations of the IUCN's Procyonid Specialist Group and the material herein is based on its findings. Of the various dangers facing the raccoon family, the most threatening is the destruction and degradation of their environments. The habitat of most of its species has been impacted over much of their range, and this has led to the decline and eradication of many of their populations. Even where they persist in degraded areas, their numbers are often reduced and isolated by increasing habitat fragmentation. The impacts on their habitats are caused by the usual litany of destructive human activities in natural areas, activities that are often interrelated: settlement expansion, increased agriculture and logging levels, wetland drainage, and overhunting.

Almost all of the procyonids inhabit forests for a substantial portion of their lives, and several of them are highly arboreal. Therefore, even a small amount of deforestation or forest disturbance can have severe negative impacts on them. Many of Central America's native forests have been ravaged, primarily for overseas markets. Even the construction of logging roads may destroy their habitats. The people who follow the logging companies into the forests are often *slash-and-burn farmers;* because the nutrients of the thin soil found in many tropical areas can only be briefly exploited, crops may only be grown for a year or two before these individuals have to move on to clear yet another site. As discussed, the red pandas of Asia are affected by many of the same problems facing their cousins in Central and South America.

International economic issues are often involved in the efforts to save habitats of procyonids and other tropical forest animals. Because the debt of many developing nations is so great, they can be pressured to further exploit their natural resources to raise money to pay off their loans. In what are known as *debt-for-nature swaps,* industrialized countries and lending agencies such as the World Bank have been urged to allow these developing nations to forego or at least reduce their debt if they engage in sustainable natural resource practices or establish new nature reserves.

Yet another problem facing the procyonids is that they are pursued over most of their range, primarily by hunters and probably to a much lesser extent by trappers. Most are sought for their fur but they are also occasionally killed for their meat. Hunters will usually profit, at least initially, from encroachments on natural areas by activities such as logging. They may form partnerships to supply game to local markets, which typically results in an even heavier toll on the wildlife. These activities are commonly unregulated or occur in areas lacking law enforcement personnel. Farmers also kill various procyonids to prevent crop raiding. Still another possible impact on their populations is caused by their young being captured for pets.

War can have grave impacts on many species, including the procyonids. Armed conflict may create widespread habitat destruction through deliberate as well as indirect activities. During a war, soldiers and civilians commonly increase their exploitation of wild animals for food and even use them for target practice. Refugees frequently hide in the forests, trying to survive by hunting and by using trees for firewood. Various examples of how warfare has led to the destruction of procyonids and their habitats can be found in El Salvador, Guatemala, and Nicaragua.

Finally, various types of pollution and their byproducts impact many species, including those in the raccoon family. An example of such pollution is acid rain, which can decrease forest cover. This effect is apparent in many areas, including the eastern United States, along coastal areas of Venezuela, near Mexico City, and in Brazil's São Paulo State.

CONSERVATION ACTION

Captive breeding programs, such as the one mentioned for the red panda, are designed to breed threatened and endangered species with the goal of releasing them back into their native habitats. If successful, these reintroductions can augment declining populations and reestablish the animals where they once occurred. Such programs may also provide important details about the basic aspects of a species' biology. They can be especially helpful in generating information about little known or elusive species such as the procyonids, several of which seem to be well suited for such projects. They usually have modest space needs and are often appealing to watch, which is a factor in attracting financial support. Many zoos have already been successful in breeding the common raccoon, and the techniques developed for these efforts might be effective in breeding the threatened island species. The possibilities for establishing captive groups of these and the other family members should be explored.

Although such breeding programs can reveal some details about a species, research on procyonids must focus on their fundamental ecological requirements. Indeed, basic information is still needed to understand the natural history of many of this family's members. Uncertainty about classifying species only adds to the difficulty of devising conservation plans for them. A clear understanding of procyonid taxonomy would help to establish conservation priorities for the potentially more unique forms.

In many areas where procyonids are likely to occur, sizable reserves have already been established, often for large Carnivores such as the jaguar. Yet most procyonids are so poorly studied that it is not known how they might be benefiting from such areas. For example, even though raccoons, coatis, and kinkajous are commonly listed as occurring in large reserves, olingos, cacomistles, and mountain coatis are rarely mentioned. Biologists should attempt to monitor the population trends of all species that are at risk and assess the impacts of habitat loss and hunting on them.

Improved public awareness is essential to the success of any conservation program. Many people may have a reasonably good appreciation of this family's members. This understanding has likely been achieved in the ways in which we ordinarily learn about wildlife species: by encountering them in their natural habitats, from visits to zoos or museums, in classes, by watching nature shows, finding out about them on the Internet or through computer programs, and perhaps by reading books such as this one. Yet, with the exception of the raccoon, this family's members are probably less well known than many other species.

Many nations have had an explosion of interest in conservation, which is evident by the greater awareness of and concern over environmental issues in their populations. Ironically, much of the increased understanding of the problems various species face occurs in areas other than those where most of these species at risk live. For procyonids to survive, the principal recommendations follow those articulated for most other species. People living around or within native forests need to understand the consequences of overexploitation of resources that occur through excessive land clearing and hunting. Public officials and other leaders must be encouraged to manage natural areas for wildlife, and create policies and practices for the sustainable use of their natural resources. Additional sizable reserves should be established and the existing ones should be better managed. Wildlife protection often needs to be strengthened through better law enforcement and additional legislation. Biologists and land use planners need to cooperate with local people in designing and administering protected areas, not just for the benefit of wildlife but for the human communities as well. Among its other advantages, increased wildlife protection can generate economic benefits, from ecotourism for example. Sadly, most of the procyonids have been so overlooked that little has been done to specifically protect either them or their habitats.

4

Form and Function

An organism's physical characteristics evolve largely in concert with its adaptation to a particular set of environmental conditions. Of course, the manner in which such traits develop is constrained by the organism's ancestry; though a coyote and a hawk might prey on the same rabbit species, they obviously employ different traits in doing so. A species' success is also dependent on various other types of characteristics, such as its physiological and behavioral traits, as well as the degree to which all of its characteristics are integrated. The raccoon's many physical attributes and the ways in which it employs them have enabled it to effectively exploit an enormous range of environments.

THE RACCOON'S PHYSICAL TRAITS

Body Size

Four basic measurements of a mammal are taken for descriptive purposes: total length, from the nose's tip to the tail's end; tail length; hind foot length; and length of the *pinna*, the ear's external portion. Most of the following measurements for adult raccoons are based on E. Raymond Hall's classic work *The Mammals of North America*. Several other sources were also used to provide a comprehensive overview of their size. As is customary, these measurements are presented in millimeters (mm), but for convenience they are also given in inches.

The body size of the raccoon, both in length and weight, varies considerably across its range. Raccoons display a moderate degree of *sexual dimorphism*, or sex-based differences in form. Males are usually 10 to 15 percent heavier than females. The total length of males ranges from 634 to 1,050 mm (25.0 to 41.3 inches) whereas that of females varies from 600 to 909 mm (23.6 to 35.8 inches). Some Maine raccoons have been reported to be 50 inches or more (1,270 mm). The tail length of males varies from 200 to 405 mm (7.9 to 15.9 inches), and those of females are 192 to 340 mm (7.6 to 13.4 inches). Male hind foot lengths are 96 to 138 mm (3.8 to 5.4 inches); those of females are 83 to 129 mm (3.3 to 5.1 inches). Their ear lengths vary from 40 to 65 mm (1.6 to 2.6 inches). Other aspects of male and female size differences are discussed in Chapter 9, "Social Organization."

Geographic body size variation in mammals generally follows a biogeographic law known as *Bergmann's Rule*, which predicts that within a species, individuals will be larger on average in the colder parts of their range. The common explanation for this pattern is that because larger individuals have less exposed surface area relative to their size (they have smaller surface-to-volume ratios), they lose proportionately less heat to the environment than do smaller ones. This pattern may also be explained by the fact that larger mammals tend to have comparatively more body fat, which has obvious insulating and food-storage benefits in colder climes. Furthermore, in the northern continents, colder environments are usually more *seasonal* than southern, warmer areas, meaning they exhibit greater seasonal variation in such factors as their temperature, precipitation, and plant productivity. Larger mammals should be more successful in these seasonal environments because they can best cope with the resource shortages that occur during the harshest times of the year.

Some evidence indicates that raccoons follow Bergmann's Rule, as the larger individuals and subspecies often occur in the more northern parts of their range. Typically, the adults vary in weight from about 3.6 to 9 kg (7.9 to 19.8 pounds), with the largest subspecies appearing to be from Idaho and nearby areas in the interior of the Pacific Northwest. Large subspecies also occur in other parts of the western and the north-central United States. As indicated, some exceptionally large raccoons come from other northern states such as Maine, where immense individuals weighing close to 13.6 kg (30 pounds) have been reported. Within one area, the average weights of adult raccoons from Illinois, Iowa, and Missouri exhibited a weight increase of about 454 g (1 pound) every 160 to

240 km (100 to 150 miles) to the north. Finally, an analysis by researchers Mark Ritke and Michael Kennedy revealed that larger raccoons also inhabit greatly seasonal areas whereas smaller ones occur in areas with the opposite conditions.

However, such patterns of body weight variation in raccoons are not at all consistent. For example, one study found that the average weight of adult males in Michigan was 6.2 kg (13.6 pounds), whereas that of the males in Missouri, much farther to the south, was 6.8 kg (14.9 pounds). The weights of the Missouri raccoons were closely correlated with the area's soil fertility, an association that an Alabama study, however, shows does not necessarily occur in other places. Furthermore, several exceptionally large individuals have been reported from the southern United States. Stanley Gehrt and Erik Fritzell conducted a comprehensive study of the raccoon population on south Texas's Welder Wildlife Refuge. Their research has provided fascinating insights into many aspects of the raccoon's ecology and behavior that are discussed throughout this book. For example, they found that the average weights of the individuals in this area were similar to those of raccoons in Michigan and other northern locales. In fact, one of the heaviest raccoons was a Texas resident that weighed 25.5 kg (56 pounds). It is just slightly lighter than the apparent all-time record holder, a Wisconsin behemoth that weighed in at 28.4 kg (62 pounds, 6 ounces). It was so hefty that the width across its back was greater than 43.2 cm (17 inches). Both were very fat males that were killed in the late autumn.

These apparent departures from Bergmann's Rule may be partially attributable to the raccoon's capacity to rapidly gain weight when food is plentiful. Another confounding factor in trying to understand their size patterns is that the smaller subspecies tend to occur both in some eastern and some southern states. This may be related to a seeming east-west or longitudinal gradient in their size variation, with the larger subspecies occurring in the west. Their size plunges in the Florida Keys, where its tiny raccoons weigh as little as 1.8 to 2.7 kg (4 to 6 pounds).

The raccoon's somewhat rotund appearance belies its considerable power and even ferocity at times. Examples abound of its almost legendary physical capabilities. One story recounts that a raccoon inside of a tree was able to support a 90.7 kg (200 pound) man who was hanging onto its tail. As those who have hunted it can attest, the raccoon possesses considerable strength for an animal of its size. It can readily defend itself against similar- or even larger-sized dogs by fighting viciously as well as intelligently. Raccoons have been reported to charge and drown dogs that were pursuing them in water. They can also display amazing endurance and resilience. Raccoons have been shaken out of trees, falling 9.1 to 12.2 m (30 to 40 feet) without apparent injury. One pregnant female is reported to have fallen at least 9.1 m (30 feet) from a tree and then successfully gave birth to her litter. Glen Sanderson, one of the foremost raccoon authorities, stated that he had never observed a raccoon injured by jumping from a tree.

Coat Color and Thickness
As is so for most mammals, the raccoon's pelt is composed of long, stiff, somewhat bristly guard hairs on the surface that help it shed moisture and a dense insulating underfur that protects it from the cold. Their fur has a grizzled look, varying from an iron gray to blackish. Their darker color is created by the black tips of the banded guard hairs; a lighter color, where it is mixed in, is produced by white tips on these hairs. Consequently, their top coat color is usually gray to black, depending on the relative number of black- versus white-tipped guard hairs.

The raccoon's coat is also normally imbued with a reddish or orange (ochraceous) color, especially on the nape of the neck and its lighter tail rings. Such coloration around the neck or shoulder area occurs in various mammals. Their throat has a distinct brownish or blackish area that is separated from the mask by thin white lines extending back from the muzzle. The predominantly dark back fur becomes thinner and grayish white or buff on the flanks and legs. On its belly, the long white guard hairs barely conceal its underfur. This part of the coat is either relatively pale, dense gray, or brownish, and is considerably more uniform in color than the coat's guard hairs. The underfur is composed of fine, short (20 to 30 mm, or 0.8 to 1.2 inches), wavy or crimped hairs, and accounts for almost 90 percent of the raccoon's coat. Males and females are similar in color, and the juveniles resemble the adults.

Like most mammals, raccoons often blend in with their environment. The color and shading of their coat appear to provide some protective camouflage. Detecting immobile individuals that are spread out, clinging to the trunks or limbs of certain trees, can be difficult. When viewed from below, its light underside may cause its outline to be blurred against a silvery night sky. As is typical of North American mammals, the lightest colored subspecies occur in the hot arid zones of the southwestern United States and northern Mexico. Darker forms tend to live in more humid areas. The darkest raccoons, which have an expanded black banding of their hair shafts, live in the humid parts of the Pacific Northwest and in tropical Central American areas with heavy rainfall. The coastal marsh–

dwelling raccoons of the southeastern United States have lighter, often reddish coats with coarser fur. Raccoons also occasionally display abnormal colorations. *Albinism*, which may cause an individual to be almost entirely white, occurs periodically, and cinnamon-colored individuals have also been reported.

The thickness of a raccoon's coat varies with latitude. Their coats tend to be heavier in the northern parts of their range, especially in the winter when they are in what the fur industry regards as *prime* condition. Raccoons in the southern United States generally have shorter guard hairs and a less dense underfur. Coat thickness may largely be a function of hair density rather than length.

Throughout their range, raccoons undergo a molt that usually begins in the spring and can last through the summer, particularly in northern populations. The molt starts on the head and undersides and then proceeds toward the back. During their molt, most of the guard hairs are lost and much of the underfur is shed. Large fur mats develop as a separate layer over their hindquarters and can remain there until late June or July. Small clumps of shed fur commonly cling to their den entrances during this time. Raccoons usually look their worst in the summer; by then their guard hairs are gone, the sparse underfur has lost its sheen, and they may have bare spots. Their new pelage is still short in the fall, but by winter, their coat has become longer and thicker. Oddly, raccoons appear to ingest more fur during their spring molt. Eating different amounts of hair may just be a function of its availability; this behavior does not appear to have any significance.

Markings

One of the raccoon's most distinguishing features is its mask. Of course, its masked appearance is associated with its well-deserved reputation of being a "bandit." This mask is a dark, fairly narrow strip of brown-black to nearly black hairs that fully surround the eye region. The one- to two-inch bands around each eye peak just above their inner corners and cover the cheeks below. The mask's striking appearance is heightened by the contrasting whitish hairs surrounding it. It usually reaches from the cheeks across the eyes and muzzle, and extends down the muzzle and up the forehead. Though relatively uncommon, some individuals have a band of faint brown hairs under the chin, making it appear as if their mask continues around the jaw.

The function of masks, which appear on various mammals, is not at all clear. They may have multiple functions within the same species and their

A standing raccoon exhibits its characteristic form and features. Photo by Glenn D. Chambers.

roles might vary from one species to the next. Facial markings such as masks probably initially evolved because they were beneficial. Through time, the value or function of these markings could have become enhanced or otherwise changed. In the least, an animal's characteristic markings such as its mask should allow it to be more recognizable, by members of its own species as well as those of others. Having distinctive markings might enable the individual raccoons in a population to recognize one another, as their facial and tail ring patterns are often unique. However, what types of cues raccoons use to identify each other is not known.

The raccoon's distinctive markings may also have a role in various other behaviors. For example, during the jaw wrestling of cubs and in clashes between adults, masks and ears may serve as targets. The masked cheeks, especially of adults, are heavily furred; thus a bite or wound in this area might not be as harmful as one inflicted elsewhere. The heads and ears of older males often appear to have been scarred by such injuries. The tail is also a target, and puncture wounds are commonly inflicted on it during brawls. Ill-fated raccoons can end up with a shortened tail or none at all. The visually accented parts of an animal's body can also function as targets during placid behaviors such as grooming. Raccoons engage in some limited mutual grooming, including facial rubbing. Occasionally, they make contact with each other on their cheeks and near or on the mask.

Their markings may also accentuate the message conveyed by certain postures. For example, when a raccoon is angry or appears threatening, it lowers its head, arches its back, elevates the shoulders, lays back its ears, and raises and thrashes its tail. Other aggressive visual displays include the baring of teeth and moving sideways in a circular motion by a series of short jumps or hops. Such movements can exaggerate an individual's size, an effect that the raccoon's markings might enhance.

According to the eminent wildlife biologist George Schaller, the raccoon's mask may also serve a purpose comparable to one proposed for the black patches surrounding the giant panda's eyes. These appear to greatly amplify the panda's small dark eyes, possibly resulting in its stare becoming more threatening. To exhibit a lack of aggressive intent, the giant panda will avert and conceal its face, covering the eye patches with its paws. A relatively small omnivore such as the raccoon should conceivably also periodically benefit from having a more threatening appearance.

The raccoon's mask might even enhance its night vision. As its almost black hairs should absorb maximal amounts of whatever night light is available, they could augment the raccoon's vision in the dark. On the other hand, the mask may also function like the charcoal that athletes smear on their cheeks to reduce glare. Because the raccoon often forages around water, a reduction of the glare caused by celestial objects, such as that from the moon reflecting off the water's surface, might enable it to hunt more effectively.

The Tail
The raccoon's tail is also striking, having five to seven brown-black rings vividly alternating with considerably lighter bands. It is fairly long— about 192 to 405 mm in adults (about 7.9 to 15.9 inches)—bushy, and cylindrical, though not prehensile like that of its cousin the kinkajou. It serves as a balance while climbing and appears to help support the body when sitting. As winter approaches in colder areas, the raccoon will store fat in its tail and may produce especially thick fat deposits just beneath the tail's skin and close to the main body. It might also attempt to shield itself from the cold by folding its tail around its feet or other extremities. During times of food shortage, which often coincide with their winter lethargy, raccoons will draw upon the fat reserves stored in various parts of their bodies, including their tails.

The Head and Skull

In addition to its coat, mask, and tail, the raccoon has several other characteristic features. Its nose is distinctively pointed. Its ears are normally prominent, being marked with both a black patch at the base and a contrasting white tip and rim. However, they may be most conspicuous when they are pulled back against the head. In this position, the black patch and white tip can be seen from the front. The sides of the muzzle, its lips, and its chin are all usually white.

Tufts of stiff whitish *vibrissae* or whiskers called *mystacials* project from the sides of its muzzle. These can range from about 50 to 100 mm (about 2 to 4 inches) in length. Most of the time, they lie against the cheeks but will stick out when an individual is alarmed, confronted, or perhaps just investigating its surroundings. They are actually enlarged white hairs supplied with sensory nerves at their bases, and they thus effectively function as tactile organs. The other four tufts of facial vibrissae are the paired *superciliaries* above the eyes, the *genal* tufts on each cheek, and a single *interramal* tuft on the chin. The raccoon also has stiff tactile hairs on the wrist's upper surface that may aid in climbing, and elongated hairs that extend beyond their nails that also appear to be sensory.

The raccoon's skull is both broad and rounded. Their *sagittal crest*, the thin, bony, ridgelike formation along the back of the skull's midsection, is highly variable in size and occasionally absent. Their *auditory bullae*, the two thin, rounded bony structures at the back of the skull's base, are laterally compressed and highly inflated on their inner sides. Like other mammals, the raccoon has a dental formula that indicates how many of a particular kind of tooth occur on the top and bottom of each side of its mouth. Their dental formula is: 3/3 incisors; 1/1 canines; 4/4 premolars; and 2/2 molars. Doubling their sum yields the total number of their teeth, which is 40. These are the same values found in both primitive procyonids and mustelids, revealing that the raccoon's dental formula has stayed the same as that of its ancestors. Its dentition is regarded as "heavy" because their broad, high-cusped cheek teeth are better suited for crushing than for slicing, as in most of the Carnivores. But despite having these adaptations for omnivory, raccoons still have the sharp, double-edged canines characteristic of their order, which they also use for predation and self-defense.

Eyes that Can Reflect

The raccoon is reported to have fairly good visual acuity. Their round eyes enable them to see rather well at night, a capacity that is facilitated

A raccoon reveals its long, stiff, whitish whiskers, also known as *mystacials*. Normally these lie against the cheek, but they will stick out when a raccoon is alarmed or investigating its surroundings. They are actually enlarged hairs that function as tactile organs. Photo by Glenn D. Chambers.

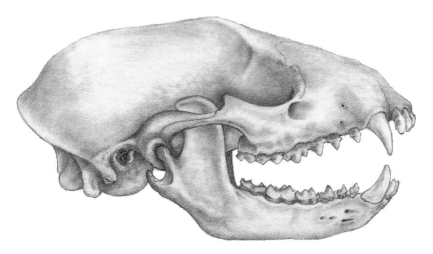

A raccoon skull. Drawing by Elizabeth Dewitte.

by the eye's shape. Vision is stimulated by light that enters the eye and is then absorbed by the retina, a sensitive light-detecting structure that lines the back of the eye and is linked to the brain via the optic nerve. The raccoon's relatively large, convex cornea allows a comparatively bright image to be projected onto its retina. Within the retina are many slender cells called *rods* that function as *photoreceptors* (light receptors). They provide sensitivity to different light intensities as well as the ability to discriminate forms and sizes. Its retina has a much smaller number of *cones*, other visual cells that determine color perception. Raccoons are probably either color-blind or perhaps "color-weak," a trait that is consistent with their nocturnal behavior. Some experiments suggest that they can distinguish colors, but they may only be able to detect differences in brightness.

Their dark eyes normally have a reddish tinge. They appear to be unusually bright, so much so that they have been described as extremely piercing, a quality that may result from the light they reflect. In many nocturnal animals, vision is enhanced by a mirrorlike reflective structure called the *tapetum lucidum* or simply the *tapetum*. Well-developed in raccoons, it is made up of layers of flattened, tilelike cells, each filled with highly refractive crystalline threads or *rodlets*. The tapetum reflects light that enters the retina, thereby magnifying stimulation of the rods. At night, the raccoon's tapetum may reflect back a large amount of light, creating a bright red eye shine. Their eyes also display a twinkling or short

Raccoon looks out while nestled in a tree. Note how the eyes are largely positioned forward, heightening the raccoon's depth perception. Photo by Julia Sims.

green flashes from certain angles, which may result from changes in the supply of blood to the tapetum.

Because their eyes are positioned somewhat forward rather than along the sides of their head, raccoons appear to have *binocular vision*. Binocular vision allows each eye's field of vision to cross, thereby enhancing *depth perception*, the ability to distinguish the spatial relations of items. However, raccoons are suspected of having poor distance vision. At least one was found to get along without any vision at all. A radiotelemetry study revealed that the activity patterns, travel routes, movement periods, and home range size of a blind raccoon differed little from one that could see.

Amazingly, it seemed to be able to move without any difficulty in its approximately 8-square-mile home range.

Ears and High-Power Hearing

Early in the twentieth century, a behavioral researcher named Lawrence Cole described hearing as a "special protective sense" for the raccoon. He recounted how the slightest sound would cause a captive raccoon to first become immobile and then so fearful that it would attempt to dash away. They reportedly can hear very faint sounds, such as that of a tiny piece of meat hitting a surface. Raccoons have a higher auditory response range than those of the dog or the red fox but a somewhat lower auditory response range (50 kHz to 85 kHz) than those of the ringtail or the coati. The higher ranges of their cousins could be related to their tendency to hunt smaller prey, which might make noises in those ranges. They may also have higher auditory ranges because these would be advantageous in the more arid environments in which they live, as sound does not carry well in such areas.

Smell and Scent Marking

Like various other Carnivores with long snouts, the raccoon's nose culminates in a shallow pad called a *rhinarium*. Theirs lacks a *philtrum*, the grooved extension that reaches from the base of the nose through the upper lip's midportion in various species. Raccoons possess an acute sense of smell; for example, they can detect acorns buried up to 2 inches under dry powdery sand. To evaluate an odor, a raccoon will occasionally pick up an item and press it against its nose. Both smell and taste seem to be involved in their food selection.

Although little about their olfactory communication is fully understood, the assessment of odors is believed to be meaningful to raccoons for several reasons. First of all, they mark their home ranges with their scent. Like other Carnivores, they have paired anal scent glands and other specialized skin glands. They deposit scent by rubbing their anal region on various items. Anal sniffing between individuals has also been observed. Urination and defecation may also be involved in their scent marking. Though they commonly deposit isolated scats, raccoons often use communal defecation sites or latrines. Some are fairly large and may be used over long periods. The more common places for these include logs, stumps, and rocks, though piles have also been found near trails, on soil mounds, and on large tree limbs. The function of these deposits is not clear, though they seem to be used in olfactory communication. Skin glands that serve in thermoregulation such as sweat glands as well as those that secrete oil for their coats presumably have odorous substances that can provide additional information. Overall, odor appraisal seems to be involved in the establishment and recognition of living areas as well as in the identification of individuals.

Paws and Touch

The raccoon has five long soft-skinned fingers on each of its hands or forefeet as well as five corresponding toes on its hind feet. Having this number of digits is referred to as being *pentadactyl*. The digits have no webbing between them, which is unusual among Carnivores. The tops of their hands and feet, including the fingers and toes, are covered by short, smooth hairs that vary from being nearly black to almost white. Their palms and hind foot soles, including the bottoms of their fingers and toes, are hairless and dark skinned. Its hind footprint closely resembles that of a human baby's foot but with longer toes. However, if it has not settled in soft soil, it often leaves an impression similar to one from the front foot: that of an adult human's hand with long, slender fingers.

Footprints of a raccoon's forefeet on the beach at Assateague Island, Virginia. All of the five fingers on each are not fully evident. Note the similarity of these prints to that of the human hand. Photo by Samuel I. Zeveloff.

Their claws are relatively short, laterally compressed, and curved. They are not retractile; as mentioned previously, those of the ringtail can partially retract. The raccoon's claws enhance its capacity for many important skills, such as climbing, ripping apart logs to search for food, and prying apart clams and oysters.

Nearly 100 years ago, Cole remarked on how touch was the raccoon's most conspicuous sense. This mammal has endured as a subject for studies of sensory and motor functions largely because of the remarkable nature of this sense. As is obvious to anyone who has observed them even briefly, raccoons are highly adept at grasping and manipulating objects. They commonly handle various objects while sitting, allowing the hind limbs to support their weight. They are exceptionally dexterous, so much so that they reportedly are able to catch flying insects. Their ability to open various types of fasteners has been well documented, and they can recall solutions to fastening problems for more than a year without practice. The manipulation of objects also appears to be a meaningful aspect of play behavior in juveniles.

The raccoon's keen powers of *tactile discrimination*, which may be thought of as the ability to differentiate objects without seeing them, have been carefully studied. They are able to distinguish between spheres with slightly different diameters, various-sized squares and cubes, and objects with dissimilar textures. They have approximately four times as many sensory receptors in their forepaw skin as they do in their hindpaw, a ratio of receptors similar to that in the human hand and foot. Actually, though, its hand is not truly comparable to that of a primate. A raccoon's hand has the typical mammalian type of skin end organs, whereas primates, including humans, possess more complex structures known as *Meissner's corpuscles* that provide even greater sensitivity when handling objects.

The part of the raccoon's brain that receives touch information from its hands, the *tactile receiving area*, is greatly enlarged relative to its size in other mammals. These brain projections, separated by grooves, correspond to specific parts of the forepaws. Such an arrangement allows information to be received from specific fingers. Raccoons also have a relatively large number of brain cells that respond to forepaw stimulation. In addition, the part of their cerebral cortex that controls forepaw motor function is considerably more developed than those in either dogs or cats. Interestingly, however, this forepaw motor function area is only about a third of the size of its forepaw sensory area. This difference is consistent with our understanding of their hand's motor capability relative to its sensitivity: that this capability, while superior to that of other Carnivores, is still considerably less well developed than their hand's sensitivity.

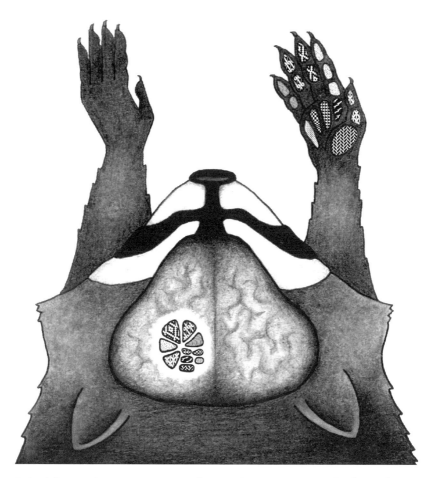

Stylized drawing showing communication between the raccoon's paw and its brain. The part of the brain that receives tactile messages is enlarged and has areas that correspond to sensory projections from distinct places on the forepaws. This arrangement should greatly enhance their tactile discriminatory abilities. Raccoons may handle their food with greater sensitivity in water as the skin on their hands becomes softened while immersed. (Note: the particular patterns on the hand and fingers here do not necessarily correspond to the analogous ones on the brain. The featured area of the brain is also not as proportionately large as that shown here.) Illustration by Elizabeth Dewitte; based on a drawing by Douglas Cramer in Radinsky 1976.

Locomotion

Although some observers suggest that the raccoon has a semiplantigrade gait, most authorities consider it to be plantigrade. As discussed in the previous chapter on the family's characteristics, semiplantigrade refers to a locomotory style in which the heels are somewhat elevated, whereas in a plantigrade gait, the entire sole makes contact with the ground. It appears that the raccoon's weight is carried slightly more on the front of its foot than on the heel, which is probably why some have described it as being semiplantigrade. In actuality, the raccoon employs different locomotory modes at various times. During routine movements, it has a slow, flat-footed walk in which it almost appears to be waddling. When in a hurry, it seems to trot. And if it needs to go extremely fast, such as when it is being pursued, it can essentially gallop. It puts both of its hind feet outside of its front feet, and takes fairly long jumps while holding its tail straight out. Generally, though, raccoons do not run. Despite being able to run for several hours or considerable distances when they are forced to, they are not especially adapted for running. This is evident when comparing them to other animals, such as the wolf, which among its other adaptations for running, has relatively long legs that produce an extended stride.

At least one researcher has been able to outrun an adult raccoon and capture it in an open area. However, if bushes or other obstacles are present, raccoons can make dodging maneuvers to avoid being caught. Though they will back down out of trees, they routinely descend headfirst. In fact, the raccoon is one of but a few mammals that can climb down a vertical tree trunk in this position. They are aided in doing this by rotating their hind foot 180 degrees at what is known as the subtalar joint; this is where the curved talus bone fits into the knobby calcaneum or heel bone. Positioning the hind feet in this direction greatly increases their traction while descending. Other procyonids can also climb down trees headfirst, but they point their hind foot backward by using different joints.

Finally, as suggested by the reference to their being able to drown dogs, raccoons are rather capable swimmers. Staying about half submerged, they swim by paddling in a manner similar to that of a dog, with their head up and tail straight out. Raccoons have an estimated swimming speed of about 4.8 km an hour (3 miles per hour). Adults have been observed to cross rivers and lakes up to 300 m wide (almost 1,000 feet or two-fifths of a mile). They regularly journey across salt water between several of the keys in south Florida. Even juveniles can cross rivers more than 150 feet wide. They will readily enter deep water to search for food such as crayfish. When pursued by hounds, they will seek sanctuary in water and can remain there for up to several hours. To dry off, a raccoon shakes itself from head to tail with a twisting, revolving action that can make it lose its balance. Impressively, such a shake will throw off a fine spray that can leave it fairly dry.

Temperature Regulation
As previously mentioned, raccoons can cool off by sweating but they also pant to regulate their temperature. Panting produces a significant flow of air that travels through the nose and then over the *rugae,* the ridges on the roof of the mouth. This air comes into contact with surfaces that are moistened with nasal and salivary gland secretions, and are well supplied with blood vessels. With each breath of dry air, the moisture can evaporate and thus cool this area's mucosal lining and the blood that flows there. Such a heat exchange also occurs between the thin-walled arteries that carry heated blood toward the head and the similar-sized veins that drain cooled blood from the nose and mouth region. This heat exchange is facilitated by a *rete,* or *rete mirabile* ("a miraculous net"), which is an aggregation of small arteries that intermesh with a group of small veins. It is part of the internal maxillary artery, a branch of the larger carotid artery. This meshwork of blood vessels reduces the temperature of the arterial blood and ultimately protects the raccoon's brain from overheating. Various types of retes in different blood vessels function similarly in many animals.

VOCAL COMMUNICATION

Communication among the individuals of a species, whether it be through physical posturing or vocalizations, facilitates the establishment of territories, aids in locating young, and is critical for mating success. Raccoons employ many types of vocalizations. During confrontations, they may scream, hiss, snarl, or growl. They may occasionally produce a series of "oinnnnggg" sounds, each at a consecutively lower pitch than the preceding one, though their meaning is not known.

Vocal exchanges between a mother and her cubs are rather complex. Their calls may account for at least 7 of the 13 types of raccoon calls that have been identified. Vocal communication is involved in all phases of their relationship. A cub's utterings can express its many needs: milk, warmth, touching, urination, and defecation. Not surprisingly, they are effective at soliciting nursing and grooming from their mother. Alternately, the mother's calls may stimulate and even synchronize suckling behavior in her cubs. When very young juveniles are upset, such as when they are separated from their mothers, they commonly make various sounds such as a high-pitched "chittering" or a "quavering purr" that has been likened to the call of a treefrog. Mothers may try to calm their young with a low "twittering" or "purring" sound, to which juveniles respond with a quiet "churr" that reflects their contentment. Researchers have described the mother's call as the "leading call" or "chitter" and the cub's response as a "whistle." Both of these vocalizations may permit individual identification, though the degree to which raccoons can do this is uncertain. Mothers and cubs may only be able to distinguish familiar versus strange sounds from these vocalizations. Their frequency depends upon the situation as well as the stage of the cub's development. When the cubs are able to move about on their own, their vocalizations decrease in frequency and intensity. At this time, their calls appear to function in reuniting them with their mother.

Vocalizations are but one element in the communication between a mother and her cubs. As suggested earlier, an individual's physical features, such as those associated with its mask, might also facilitate recognition. Calls, however, may be more reliable to raccoons for detection than visual cues because they are usually most active at night, often in forests. Though chemical signals may also be used to recognize or locate one another, they tend to work slowly and normally do not provide information on direction as well as visual or vocal cues. Whereas a mother may take several minutes to locate her cubs from chemical cues, such as by sniffing their tracks, they should be able to find one another more rapidly using vocalizations.

Evidently, the raccoon's various sensory organs and its vocalizations are highly developed. Many observers believe that it has a relatively high level of intelligence. Cole suggested that raccoons were intermediate between monkeys (though the species was not mentioned) and cats in the speed with which they form associations, and even closer to monkeys in the complexity of the associations they can make. It is curious that behavioral researchers have relatively ignored the raccoon given its great adaptability and apparently keen intelligence. In fact, as discussed in the following chapter, the raccoon's traits have allowed it to become one of the most adaptable and wide-ranging mammals in North and Central America.

5

Distribution
and
Subspecies

Raccoons have a vast transcontinental distribution, occurring throughout most of North America and Central America. They are found from across southern Canada all the way to Panama, as well as on islands near coastal areas. They occur in each of the 49 states of the continental United States. In just the past few decades, they have significantly extended their range in several areas, including the Rocky Mountain West of the United States, and the Canadian provinces of Alberta, Saskatchewan, and Ontario. About 50 years ago, their northernmost locale was Vancouver Island in Canada's British Columbia. Now they presumably occur further to the north in several of Canada's south-central provinces. Although raccoons are native only to the Western Hemisphere, they have been successfully transplanted to other parts of the globe.

THE RACCOON'S SUCCESS

Within approximately the last 70 years, the raccoon has seen a significant reversal of its fortunes, both in its numbers and distribution. Following a decline to a relatively low level in the 1930s, they have since dramatically rebounded to their present healthy position. They began to prosper following their 1943 breeding season, apparently across the continent. A rapid population surge continued throughout the 1940s, and high numbers have been sustained ever since. By the late 1980s, the number of raccoons in North America was estimated to be at least 15 to 20 times the amount that existed during the 1930s. The reasons for both their decline and subsequent increase are not well understood, though several explanations for their proliferation are explored in the following discussion.

By now, their numbers have undoubtedly grown even more, as they have continued to expand into habitats where they were once either rare or absent, such as sandy prairies, deserts, coastal marshes, and mountains. Their spread throughout the Rocky Mountain West is indicative of the fast pace at which they can exploit new environments. Scarce or absent west of the Continental Divide until the 1960s, they are now abundant in most of Colorado. Similarly, in 1965 they were only found in the eastern one-fifth of Wyoming, but by 1989 occurred in all but 2 of its 23 counties. They likely spread into western and central Wyoming by traveling from northeastern Wyoming and Montana into the Bighorn and Wind River basins. They have expanded so far north into Canada so quickly that as of the late 1980s, the natives in some areas still had no name for it. Despite significant numbers being harvested and having suffered occasional declines, typically because of disease, the raccoon has consistently maintained high population levels.

Several factors explain the raccoon's dramatic increase in abundance and distribution. First, their success has been partially attributed to the growth of cities, as they often thrive in suburban and even urban settings. Furthermore, they have been deliberately introduced throughout the continent. Within the United States, they are commonly taken from one area to another, both legally and illegally, to restock hunting areas and presumably because people simply want them to be a part of their local fauna. Their appearance and subsequent flourishing in Utah's Great Salt Lake Valley within the last 40 years appears to be from such an introduction. As an example of the ease with which transplanted individuals can succeed, raccoons from Indiana have reportedly been able to flourish on islands off the coast of Alaska.

The raccoon's expansion in various areas is likely due to the spread of agriculture, especially the corn-growing industry. It has evidently benefited from the European colonization of the prairie pothole region of south-central Canada and the north-central United States. Glaciated depressions (*potholes*) that capture rain or melted snow have created numerous wetlands in this area. The region extends from Alberta, Saskatchewan, and Manitoba across northeastern Montana, then southeast through North Dakota and eastern South Dakota into western Minnesota and northwestern and central Iowa. Raccoons have been able to exploit this area's crops, such as corn and cereal grain, which have become dependable food sources for them. Raccoons and several other mammals have also taken advantage of the presence of corn and other crops in the northeastern United States. In some parts of this region, their populations have multiplied more than tenfold within approximately the last 20 years.

The expansion of agriculture, however, does not necessarily lead to rapid increases in their abundance. For example, farming in Kansas and eastern Colorado proceeded rapidly in the 1870s and 1880s, but this was about 50 years before raccoons started to spread out from the wooded river bottomlands, their major habitat in that region. They have also expanded into many areas lacking any agriculture other than grazing as well as into places without forests or permanent streams.

Prior to the Europeans settling the prairie pothole region, raccoons probably just occurred along its rivers and streams and in the wooded areas of its southeastern section. With the possible exception of southern Manitoba, they were absent from Canada. They first became more widely distributed in the southern part of this region, and by the 1940s were abundant throughout its southeastern portion. In the 1950s, their populations swelled in Canada. The control of coyotes in the prairie pothole region in the 1950s may have been a factor in this area's raccoon expansion. If their numbers are sufficient, coyotes might be able to suppress raccoon populations (though little direct evidence supports this notion). By the 1960s, the raccoon had become a major predator of the canvasback ducks nesting in southwestern Manitoba.

The extermination of the wolf from most of the contiguous United States may have been a critical factor in the raccoon's expansion and numerical increase. In colonial times, when the wolf's range included almost all of North America, raccoons apparently were only abundant in the deciduous forests of the East, Gulf Coast, and Great Lakes regions, though they also extended into the wooded bottomlands of the Midwest's major rivers. In such areas, their arboreal habits and the presence of hollow den trees should have offered some protection from wolves and other large predators. Even though raccoons may not have been a significant part of their diet, wolves surely would have tried to prey on those exposed in relatively treeless areas.

Between 1915 and 1940, concerted efforts finally eliminated the wolf from the lower 48 states, except for a few remote places in the western mountains and other wilderness areas. Surveys indicate that raccoons were abundant in some deciduous forests and riparian woodlands in the Rocky Mountains and Plains states, but that they were scarce in arid, treeless habitats until around the 1940s. At that time, wolf populations greatly declined and the raccoon began to spread into more open country. Now raccoons occur in almost every part of the Rocky Mountain region.

Despite considerable opposition, wolves have recently been reintroduced into several portions of their former range, such as Yellowstone National Park, and will likely spread into adjacent areas and perhaps beyond. There are also plans to reintroduce other Carnivores, such as the Canada lynx and the grizzly bear, into parts of their former ranges. These reintroductions might affect the raccoon's recent expansion.

Unfortunately, early records of the raccoon from North America's interior as well as those of its subsequent spread are scarce. The pioneering mammalogist Vernon Bailey (1926:188) provided a useful review of some initial reports on the raccoon: "On the Missouri River no mention is made of raccoons by Lewis and Clark [1806], Maximilian [1833–34], or Audubon [1843], while Hayden [1855–57] . . . reports them abundant at Council Bluffs [Iowa]; but the highest point on the Missouri River at which he observed them was about the mouth of the Niobara River [in Nebraska]" (bracketed insertions follow Finley 1995). In 1877, Wooden Leg, a Northern Cheyenne warrior, was transported to the Southern Cheyenne reservation in western Oklahoma. While there, he shot a raccoon but did not know what it was. Born in the Black Hills of the Dakotas, he had roamed extensively on the high plains of southeastern Montana and northeastern Wyoming, as well as throughout western Nebraska and Kansas. Given his likely knowledge of the area's fauna, Wooden Leg's lack of familiarity with raccoons indicates that they did not occur on the northern Great Plains until close to 1877. In 1913, raccoon tracks were observed along the Missouri River as far as Fort Clark, North Dakota. By 1974, they were considered to be common, though recent, inhabitants of the Black Hills.

Raccoons also occur in many parts of Eurasia, though this is not within their natural range. They have been introduced both deliberately and as a result of the escape of fur farm captives in France, Germany, and the former Soviet Union. Introduced into Germany in 1934, their population reached 4,000 to 5,000 individuals in 30 years. By the 1970s, they occurred in a considerable portion of what was then West Germany and were said to pose a serious threat to its wildlife. From Germany, they have spread into France and The Netherlands. There have also been reports of raccoons, mostly isolated individuals, in Great Britain. Raccoons were released in the former Soviet Union in 1936 and their commercial fur trapping began by 1955. By 1964, they had swelled to an estimated population of 40,000 to 45,000. Nevertheless, by around 1980, stable populations may have only existed in the Caucasus region and Byelorussia. Raccoons may be unable to thrive in areas such as those in the former Soviet Union, where the winters produce deep, loosely packed snow.

RACCOON SUBSPECIES

Species with extensive distributions are typically composed of numerous varieties, and the raccoon is certainly no exception. The individuals belonging to these varieties, also regarded as *races* or subspecies, can interbreed and thus belong to the same species. Their variations, however, can be meaningful as they may connote adaptations to particular ecological circumstances. As stated, the raccoon's size often varies geographically, with the largest subspecies ordinarily found in more northern, seasonal climes. Some subspecies also exhibit variation in the development of the mask and other facial markings. Geographic variation in their coat color is largely restricted to differences in its general tone. As also mentioned, the paler subspecies usually occur in desertlike areas such as the Colorado River Valley whereas darker forms are found in such regions as the Pacific Northwest. Most of the raccoon subspecies, however, are hard to distinguish unless their localities are known. Only a few investigations have explored genetic variability in raccoons. One found significant differences between individuals from the northwestern and eastern United States, indicating that important genetic distinctions may occur between at least some of their subspecies.

The following descriptions of the 25 common raccoon subspecies are largely based on Edward A. Goldman's definitive work *Raccoons of North and Middle America*, a volume in the U.S. Fish and Wildlife Service's authoritative North American Fauna series. Since this volume was published, the distributions of several subspecies have changed, significantly in some cases. Much of the recent range information is based on material in Hall's *The Mammals of North America* as well as scientific journal articles.

The Eastern Raccoon (*Procyon lotor lotor*)
This subspecies occurs from Nova Scotia, New Brunswick, southern Quebec, and southern Ontario, south through the eastern United States to North Carolina, and from the Atlantic Coast west to Lake Michigan, Indiana, southern Illinois, western Kentucky, and presumably eastern Tennessee. It is a comparatively small, dark raccoon with a long, full coat. Its upper parts are a buffy gray overlaid with black; its sides are lighter. The top of its head has a coarsely grizzled appearance. Its mask is bordered

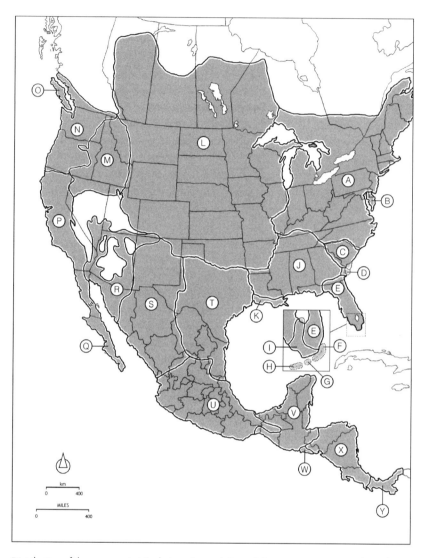

Distribution of the raccoon in North America and Central America. Approximate boundaries for its 25 subspecies are indicated: A. Eastern raccoon, B. Coastal Marsh raccoon, C. Hilton Head Island raccoon, D. Saint Simon Island raccoon, E. Florida raccoon, F. Matecumbe Key raccoon, G. Key Vaca raccoon, H. Torch Key raccoon, I. Ten Thousand Islands raccoon, J. Alabama raccoon, K. Mississippi Delta raccoon, L. Upper Mississippi Valley raccoon, M. Snake River Valley raccoon, N. Pacific Northwest raccoon, O. Vancouver Island raccoon, P. California raccoon, Q. Baja California raccoon (Note: The uncertain boundary between the California and Baja California raccoons' distributions is indicated by interrupted lines), R. Colorado Desert raccoon, S. Mexican raccoon, T. Texas raccoon, U. Mexican Plateau raccoon, V. Campeche raccoon, W. Salvador raccoon, X. Costa Rican raccoon, Y. Isthmian raccoon. Illustration by Elizabeth Dewitte; based on information from Findley 1995, Hall 1981, and Sanderson 1987.

by broad, whitish lines, and it is often discontinuous in the middle where a blackish median line is largely isolated by lighter ones. Its skull is relatively small. Although its characteristics are fairly consistent, some of them overlap with those of four other subspecies. It is similar to the Upper Mississippi Valley raccoon of Minnesota, but it is much smaller. It has a darker, less grayish cast than the otherwise comparable Hilton Head Island raccoon. It also resembles the Saint Simon Island raccoon but has a longer and softer pelage. Last, though it is also similar to the Alabama raccoon, the Eastern raccoon is typically larger, having a bigger skull and heavier overall proportions, and has a darker, considerably longer pelage.

The Coastal Marsh Raccoon (*Procyon lotor maritimus*)
This subspecies occurs in the marshes of the Delmarva Peninsula, a land form so named because it extends through sections of Delaware, Maryland, and Virginia on the Chesapeake Bay's eastern side. Because it is both smaller and paler than the Eastern raccoon, telling them apart is easy. Usually, the Coastal Marsh raccoon's upper parts are a pale buffy gray. Perhaps its most distinguishing trait, at least among the raccoons on the East Coast, is its very long and coarse hairs.

The Hilton Head Island Raccoon (*Procyon lotor solutus*)
This small raccoon lives along the coastal strip and islands of South Carolina. Despite its name, it also occurs inland, perhaps throughout much of the state. Its long coat is particularly dense in the winter. It is grayish though considerably overlaid with black, especially along the midback area. Its black mask is normally unbroken, and its small nape patch is bathed with a yellowish buff color. It is generally a more solid gray than the Eastern raccoon. Though similar to its cousin to the south, the Saint Simon Island raccoon, its coat is a more distinct gray and less likely to appear buff or brown. It also has a considerably lighter dentition than the Saint Simon Island raccoon.

The Saint Simon Island Raccoon (*Procyon lotor litoreus*)
This raccoon lives along the coastal areas and islands of Georgia. Dark and medium-sized, it is similar to the Florida raccoon. It is more buff or brown than South Carolina's Hilton Head Island raccoon. Though similar in color to those in some Eastern raccoon populations in Pennsylvania, it has a shorter, more bristly coat. Interestingly, its remarkably heavy dentition can distinguish this raccoon from the other eastern United States subspecies.

The Florida Raccoon (*Procyon lotor elucus*)
This subspecies occurs in peninsular Florida and extreme southern Georgia, and has been introduced to Grand Bahama Island. In southwestern Florida it is replaced by the Ten Thousand Islands raccoon, and in northwestern Florida, it *grades* (gradually merges with another subspecies through a series of intermediate forms) into the Alabama raccoon. The Florida raccoon is medium sized and generally dark colored. Its *nuchal* patch, that area of its coat on the nape of the neck has a rich, deep rusty color. Rather similar in appearance to the Saint Simon Island raccoon, it may be best distinguished from that subspecies by its much lighter dentition. It also looks like the Eastern raccoon but is usually paler. The frontal sinuses of its skull are notably inflated, giving it a humped appearance.

The Matecumbe Key Raccoon (*Procyon lotor inesperatus*)
This subspecies lives in the greater Key Largo island group along Florida's southeast coast from Virginia Key just off of Miami, south to Lower Matecumbe Key. Though closely allied with the Florida raccoon of the adjacent mainland, it is usually smaller and grayer, especially on the head. It also has a more restricted mask, and its snout's upper surface is paler. The frontal area of its skull is markedly depressed, appearing flat rather than humped. An extensive road network between Miami and Key West links the mainland to the keys and many of the keys to each other. Undoubtedly, this has enabled the various subspecies there to interbreed, and the traits that once distinguished them have probably become considerably diminished during the past several decades.

The Key Vaca Raccoon (*Procyon lotor auspicatus*)
This raccoon has the dubious distinction of having the narrowest range of any of the Florida subspecies. As its name implies, it lives on Key Vaca; it also occurs on the closely adjacent keys. This area is the central section of the main chain of keys off of the Florida coast. This raccoon is very small and pale. Though similar in size to the Ten Thousand Islands and Torch Key raccoons, it is clearly paler than the former subspecies and less pale than the latter. As well, its cranium differs from those in both of the other raccoons. It is also smaller and paler than the Matecumbe Key raccoon on Upper Matecumbe Key. Its upper parts are typically thinly overlaid with a rusty brown hue.

The Torch Key Raccoon (*Procyon lotor incautus*)
This is the race of Florida raccoons that lives furthest from the mainland, on and in between No Name Key and Key West on what is known as the

Big Pine Key group. A small subspecies, it is the palest of all of the Florida forms, a trait that may have been selected for by the region's brilliant light. The skull's highly arched nature resembles that of the Florida raccoon. Though usually paler, especially on the head and face, it is close in color to the Key Vaca raccoon. Its mask, though, is more restricted and is distinctly interrupted between the eyes. It is clearly paler than both the Ten Thousand Islands raccoon and the Matecumbe Key raccoon on Upper Matecumbe Key. Various cranial traits also distinguish it from the latter three subspecies. Like the other raccoons on the Florida Keys, the Torch Key raccoon resides mainly, and sometimes entirely, in mangrove swamps without any fresh water except that produced by rain.

The Ten Thousand Islands Raccoon (*Procyon lotor marinus*)
This raccoon is probably limited to the mangrove-covered and -bordered keys in southwestern Florida's Ten Thousand Islands group. It also occurs on the adjoining mainland from Cape Sable north through the Everglades to Lake Okeechobee. It fittingly inhabits a place named Coon Key. At high tide, most of these islets become swamped by three to four feet of water and some may lack any fresh water. Furthermore, most have no dry land and lack potential den trees. Thus, the raccoons may make their homes on the large exposed mangrove roots, but they can be forced to retreat from these once the tide comes in. Although one of the smallest subspecies, it possesses a heavy dentition. Its color resembles that of the Florida and Matecumbe Key raccoons, however it is usually grayer, particularly on the head. It is also typically smaller than these subspecies. Moreover, its back is less heavily overlaid with black and its rusty nape patch is not as distinctive as those of the other forms. Its mask is also more restricted than that of the Florida raccoon. Finally, the Ten Thousand Islands raccoon has a much smaller, more delicately proportioned skull than the Florida raccoon, as well as a more rounded braincase with a considerably more depressed frontal area.

The Alabama Raccoon (*Procyon lotor varius*)
This is a small subspecies that ranges across far southwestern Kentucky, Tennessee, Mississippi, northern Louisiana, Alabama, northwestern Florida, and western Georgia. It has a small and slender skull. The coat's upper areas are a light buffy gray, thinly covered with black. Along the middle of its back, it is imbued with a light yellowish buff color that is heightened on the shoulders and neck. Its sides are a more distinct gray. Its mask is brownish black changing to a rusty brownish between the eyes. Although this raccoon most closely resembles the Eastern raccoon, it is smaller,

generally paler, and has a considerably shorter coat. It is similar in color to the Texas raccoon, but again is much smaller. Even though the Alabama raccoon grades into the Florida, Texas, and Saint Simon Island raccoons, the distinction between each of them is rather clear.

The Mississippi Delta Raccoon (*Procyon lotor megalodous*)
This raccoon resides in southern Louisiana's coastal region and on several nearby islands. It can be readily differentiated from the other subspecies by its distinctive yellowish coloration. Its imposing skull has a characteristic frontal hump and it has a heavy dentition, meaning that its teeth are proportionately large relative to those of similar-sized raccoon subspecies.

The Upper Mississippi Valley Raccoon (*Procyon lotor hirtus*)
This large subspecies ranges through the upper Mississippi and Missouri River drainages, throughout Montana, Wyoming, and Colorado all the way to Lake Michigan, and from the south-central Canadian provinces down to southern Oklahoma and Arkansas. Within the last few decades, this already widely distributed raccoon has embarked upon an impressive range expansion into far northern Alberta and throughout several states in the western United States. The raccoons now in the Green River drainage are presumed to be from this subspecies. In addition, this is likely the subspecies that ranges across Wyoming, with the possible exception of those in the Snake River drainage, which may be Snake River Valley raccoons. The Upper Mississippi Valley raccoon is dark and has a long and full soft coat that is normally suffused with a buff color. Though similar to the Eastern raccoon, it is ordinarily much larger and has a larger skull and a longer, more buff-colored coat. Though it is also comparable in size to the Texas raccoon, the Upper Mississippi Valley raccoon is larger and has a longer, denser coat.

The Snake River Valley Raccoon (*Procyon lotor excelsus*)
This subspecies lives in the Snake River drainage area of southeastern Washington, eastern Oregon, and Idaho, from where it likely spread into the river's Wyoming drainage. It also occurs in the Humboldt River Valley of Nevada and in the corner of northeastern California. It has a massive skull and is reputed to be the largest raccoon subspecies. Its upper parts are a light buffy gray, moderately covered with black. The sides are a more distinct gray and its head is grizzled black and gray. Although it is closely related to the Pacific Northwest raccoon, it is considerably

larger and paler and has a much grayer head. Similarly, it is usually larger and paler than the California raccoon.

The Pacific Northwest Raccoon (*Procyon lotor pacificus*)

This subspecies occurs in southwestern British Columbia, except on Vancouver Island; Washington, except in its southeastern portion; the western two-thirds of Oregon; and extreme northwestern California. It is regarded as *the* raccoon of the Pacific Northwest coastal and Cascade Range regions. A dark, cinnamon brown subspecies, its upper body is extensively covered with long black hairs. It closely resembles those in Sacramento Valley populations of the California raccoon, but is darker and has a shorter, wider skull. It is smaller and darker than those in Snake River Valley raccoon populations in southeastern Oregon, and it is distinctly larger than the Vancouver Island raccoon.

The Vancouver Island Raccoon (*Procyon lotor vancouverensis*)

This subspecies of raccoon resides primarily on Vancouver Island. It also occurs on several nearby minor islands, including Saltspring, Pender, and Saturna, and has been introduced on Cox Island, Graham Island, and the Scott Island group of British Columbia. It is rather dark; its generally gray upper parts are heavily overlaid with black and its head is also mainly black. It most closely resembles the Pacific Northwest raccoon but is considerably smaller. Because it is Vancouver Island's sole subspecies and it is largely restricted to this island, it is clearly recognizable.

The California Raccoon (*Procyon lotor psora*)

This raccoon lives throughout most of the state from which it received its name, except in California's extreme northwest coastal section, the three-state border area at its northeastern corner, and its southeastern desert region. It occurs south through northwestern Baja California and into far west-central Nevada. It is a large, moderately dark subspecies with a broad, flat skull. Though very similar to those in populations of the Pacific Northwest raccoon in Washington, it is normally paler. It is also darker and a less ash gray than the Colorado Desert raccoon. The skulls of these two subspecies are similar in size, but the California raccoon's is broader and has a flatter frontal region. This raccoon is also smaller and usually darker than the Snake River Valley raccoon. Whereas it looks somewhat like the Eastern raccoon, its upper parts are grayer and less suffused with a buff color. And as is typical of western forms, their black mask is continuous; the Eastern raccoon's mask is distinctly interrupted on either side of its median line.

The Baja California Raccoon (*Procyon lotor grinnelli*)

This subspecies occupies the southern half of Baja California, from the Cape region north to San Ignacio. It is a large, pale variety with a broad, highly arched skull. Its upper parts are usually a coarsely grizzled iron gray. The middle of its back is just faintly suffused with a pale buff hue that becomes more distinct on top of its neck. Rather thinly overlaid with black, the top of its head is a gray mixed with black, again producing a grizzled effect. It has a continuous black mask. Though similar to the Colorado Desert raccoon, it is slightly darker and its skull has a more evenly arched profile. It is generally paler and grayer than those in Sacramento Valley populations of the California raccoon. Given the extreme aridity of its environment and its characteristic dependence on water, it likely has a rather interrupted distribution.

The Colorado Desert Raccoon (*Procyon lotor pallidus*)

This raccoon occurs in the Colorado and Gila River Valleys, a considerable segment of south-central Arizona, the adjoining areas north to northeastern Utah, and east to western Colorado and northwestern New Mexico. Since about 1970, it has extended its range up the Colorado and Green River canyons into the Colorado Plateau region, and the Yampa Valley of northwestern Colorado. It closely resembles the Mexican raccoon both in size and color; however, their separation seems valid because of their distinct skull characteristics. Though its color varies considerably, it is much paler and a more ash gray than the California raccoons in Sacramento Valley and is slightly paler than the Baja California raccoon. Finally, the Colorado Desert raccoon commonly has seven to eight narrow black tail bands, whereas another Rocky Mountain subspecies, the Upper Mississippi Valley raccoon, typically has wider and fewer than seven tail bands.

The Mexican Raccoon (*Procyon lotor mexicanus*)

This subspecies occurs in New Mexico except in its northeastern and northwestern parts, southeastern Arizona, western Texas, and then south through the Mexican provinces of Chihuahua, eastern Sonora, Sinaloa, and Durango to northern Nayarit. Along with the Colorado Desert raccoon, it is one of the palest subspecies. Externally, these two races are virtually indistinguishable; however, the Mexican raccoon's skull is usually broader. It is paler than such similar subspecies as the Mexican Plateau and Texas raccoons. Its upper parts are typically a coarsely grizzled iron gray and its underparts are a light buff. They have a broad uninterrupted facial mask.

The Texas Raccoon (*Procyon lotor fuscipes*)

This subspecies resides in most of Texas, except for its extreme northern and western areas. It also occurs in southern Arkansas; Louisiana, except for the Mississippi Delta region; and northeastern Mexico, including Coahuila and Nuevo Leon to southern Tamaulipas. It is a large, dark grayish raccoon with a medium-length pelage. One of its distinguishing traits, relative to the raccoons in surrounding regions, is its uniformly black and continuous mask. Though similar in size to those in some Minnesota populations of the Upper Mississippi Valley raccoon, it is grayer and less suffused with buff and has a shorter, less-dense pelage. It is decidedly darker than those in Chihuahua populations of the Mexican raccoon, and less grayish than those in Valley of Mexico populations of the Mexican Plateau raccoon. Though similar in color to the Alabama raccoon, it is usually grayer and much larger.

The Mexican Plateau Raccoon (*Procyon lotor hernandezii*)

This raccoon occurs throughout a considerable segment of southern Mexico's plateau region and its adjoining coasts—from Nayarit, Jalisco, and San Luis Potosí, and south into the Isthmus of Tehuantepec, with a fingerlike range projection south of the Yucatán Peninsula's western edge. It is a large grayish subspecies with a heavy dentition. While similar to the Texas raccoon, it is grayer above and has a flatter skull. It is also clearly darker than the Mexican raccoon in Chihuahua, with which it intergrades in western Mexico. Compared to the Campeche raccoon, it has a longer coat that is more heavily overlaid with black.

The Campeche Raccoon (*Procyon lotor shufeldti*)

This subspecies ranges in Mexico from the Isthmus of Tehuantepec east through Chiapas, Tabasco, and throughout the Yucatán Peninsula including Campeche, into Belize and Guatemala, to western Honduras. It is large, short haired, and pale. Apparently, its members are darker where they intergrade with Central American subspecies. Its upper parts are usually a light buffy gray, covered with thinly distributed black-tipped hairs, resulting in a coarsely grizzled look. The top of its head is a distinct gray though mixed with black. Its mask is substantial, extending downward along the midline of its muzzle to its nose and upward to the middle of the forehead. Compared to its close relative, the Mexican Plateau raccoon, it is paler above, especially on its head. Though their skulls are similar in size, the Campeche raccoon's is more massive.

The Salvador Raccoon (*Procyon lotor dickeyi*)

This is the most northern of the Central American raccoons. It appears to be highly restricted to mangroves and has but a small distribution in the coastal region of southwestern El Salvador (though it probably also occurs in southeastern Guatemala). One of the darkest subspecies, it is medium sized and has a short, thin-walled skull. Their upper parts are grayish and heavily overlaid with a black cast that extends well down its sides. Atop the head, it is a distinct gray but still heavily mixed with black, producing a grizzled effect. It has an extensive facial mask. Their large cheek teeth are often surprisingly worn. Such wear, as well as their fragile skulls, are indicative of an inadequate diet. Their excessive molar wear may also be due to a heavy subsistence on hard-shelled crabs. It is similar in size and color to the Costa Rican raccoon, but it differs in several traits, including those of its unusual skull.

The Costa Rican Raccoon (*Procyon lotor crassidens*)

This raccoon resides in Costa Rica; Nicaragua; El Salvador, except for its southwestern coastal region; Honduras; and western Panama. Another of the darkest raccoons, its back is heavily overlaid with black extending down the sides. Though it externally resembles both the Isthmian and Salvador raccoons, its skull is rather distinctive. It is larger than the Isthmian raccoon's and more massive than the Salvador raccoon's.

The Isthmian Raccoon (*Procyon lotor pumilus*)

This subspecies occurs in Panama, from Porto Bello west to Boqueron, Chiriquí. Its range marks the southern limit for the entire species, and it overlaps with that of the crab-eating raccoon. It also eats crabs, though the degree to which it relies on them is unknown. Closely allied with the Costa Rican raccoon, it is also very dark as its upper coat is heavily overlaid with black. The Isthmian raccoon, however, has a smaller, shorter, and broader skull.

Many of these subspecies, especially those residing in neighboring habitats, are difficult to distinguish from each other. Given the normal variation in appearance that occurs within a subspecies, an individual of one may easily be mistaken for that of another. Moreover, several of these subspecies can only be identified after a close inspection of their features, particularly their skulls. Thus, even a competent naturalist might not be able to distinguish the raccoon subspecies using detailed descriptions, especially those from the boundaries of their distributions. With so many subspecies occurring over such an immense area, it should not be at all surprising that the raccoon lives in a great variety of habitats under many different conditions.

6

Living
Arrangements

An animal's habitat often defines its ecology, or how it adapts to and persists within its environment. As one that prefers wooded and wet areas, the raccoon often uses trees for dens and protection and aquatic animals for food. However, its adaptability has allowed it to make the most of almost any habitat and what it may have to offer, even as this may change from season to season or day to day. This malleable animal also modifies its activity levels to adjust to a particular clime, as evidenced by its winter lethargy and the annual fluctuations in its weight.

HABITATS

Because raccoons can thrive in an enormous variety of habitats, they have one of the widest distributions of any North American mammal. Normally, they are most abundant near water, traveling along the shores of streams and lakes in search of food. They tend to be most numerous in *mesic* or moderately moist habitats such as hardwood and mangrove swamps, fresh and saltwater marshes, and bottomland forests. The latter areas, also known as floodplain forests, are usually lower in elevation than the surrounding riverbanks. Raccoons are also common in suburban residential areas, as well as on cultivated and abandoned farmlands. These

Raccoon at the water's edge, likely searching for food. Photo by Glenn D. Chambers.

habitats are only where they are likeliest to be abundant; high raccoon densities can occur in many environments.

Despite their proclivity for wet habitats, within the last 60 years raccoons have prospered in countless other settings. During their continent-wide expansion in the 1940s, they began to move into areas where both water and trees were scarce, such as plains and deserts. However, they generally have low densities in deserts, where they are restricted to stream environments. Raccoons avoid large open fields and pastures, so as they expanded their range onto the prairies of the northern United States and southern Canada, they sought out buildings, wooded lots, and wetlands. They have also benefited from certain modified landscape features. For example, on the prairie farmlands of the central United States, the ditch banks not grazed by livestock may serve as raccoon hiding places or travel routes. Nevertheless, in such atypical habitats they still seem to fare better under relatively familiar conditions. Thus, the woody vegetation near permanent water sources on the grasslands of the central states tends to be important to them. Their numbers are also usually low in dry upland forests, especially where pines are mixed with hardwoods, as well as in more pure pine forests such as in several southern states. Last, they rarely occur in mountains that are higher than 2,000 m (6,560 feet).

Throughout their range, raccoons use a fantastic assortment of natural and human-made structures for their winter dens and as sites for rearing offspring, including the ground dens of other animals; rubbish, brush, and lumber piles; corn cribs; sheds; haylofts; and, much to the consternation of home owners, attics, chimneys, and wall spaces. As a result, they can also prosper in relatively treeless areas. Similarly, in places lacking naturally occurring water or where it is in short supply, they can survive by using water intended for livestock and irrigation, or by taking advantage of birdbaths and even swimming pools.

Although they have success in such varied settings, raccoons still appear to depend upon particular habitat features to reach high and stable population levels. Early studies in the midwestern states in the 1940s stressed their reliance on trees and water. In Ohio, for example, their substantial population decline in the decades leading up to that era was attributed to several factors: a 50 percent reduction in the length of permanent streams during the previous 75 years, a loss of forests and thus potential den sites, and a decrease in swamp and bog habitats (*bogs* are areas with wet, spongy ground that are usually rich in accumulated plant matter). Such findings confirm that waterways, wetland habitats, and wooded areas are all vital components of good raccoon habitat. Their strong affiliation with forests is further exemplified by their tendency to

Raccoon in the eaves of a barn. These animals use a great variety of human-made structures for their dens. Photo by Rachelle Hansen.

remain in trees during flooding rather than escape to higher ground as do many other animals.

John Pedlar and his colleagues described in a recent study how raccoons can also benefit from certain habitat features in agricultural settings in eastern Ontario, Canada. The individuals in this study that were within and around agricultural areas were found to be more active near *fencerows*, where vegetation grows around the borders between fields. Fencerows often include fruit-bearing plants such as raspberries, have an abundance of insects and small mammals, and provide cover. The raccoons in this region were also more evident around *edge* areas, the borders between habitats, and may be responsible for the high levels of bird nest predation that occur in those areas. In this study, the woods near farms furnished the raccoons with both dens and protective cover. Because crops essentially surrounded these wooded refuges, they had the potential to support large raccoon populations. Similarly, raccoons in urban areas can use any remaining wooded tracts for daytime cover, and then head to the city to feast upon prey and waste material at night.

DENS AND RESTING SITES

Raccoons use shelters for several purposes. For their often extended winter "sleep," the most commonly used dens are hollow tree trunks or, if they are large enough, hollow tree limbs. This period of lethargy is described in the section "Activity and Inactivity," which follows later in this chapter. In a Michigan study, their den cavities averaged 29 by 36 cm (about 11.5 by 14 inches) and were usually between 3 to 12 m (about 10 to 42.6 feet) above the ground. As long as they can remain dry, their dens do not have to be high above the ground. Apparently, raccoons do not add nesting material to their dens, though any decayed woody material that accumulates as the cavity forms should make them more comfortable. In Michigan, they used a variety of hardwood trees for dens, but especially favored red maple. Beech was not used, presumably because its smooth bark makes climbing difficult. In another study in southwestern Virginia's mountains, the most common tree cover types were swamp chestnut oak and tupelo gum–red maple stands. Whereas their nightly feeding forays occurred in both forest zones, their denning was almost entirely confined to the latter areas.

Not surprisingly, trees are less critical to raccoons in the warmer parts of their range. In their colder haunts though, tree cavities can provide an effective buffer against temperature extremes. In Kansas, for example, their tree cavity temperatures ranged from −4 to 24°C (about 25 to 75°F) during a period when the outside temperature varied from −8 to 28.5°C (about 17.5 to 83°F). Furthermore, when a raccoon enters its den, the cavity's temperature immediately rises from the partial containment of its body heat. The temperatures inside the den are also more stable than those outside; a drop in the outside temperature can be delayed by as much as two hours in a den. Last, the cavity's temperature changes may be less abrupt. In Kansas, while the outside temperature fell by as much as 12°C (23°F) in an hour, the den temperature usually dropped by no more than 1°C (2°F) in that time.

Although raccoon densities are usually highest where tree dens are plentiful, ground dens can be critical to their success, especially where hollow trees are scarce. They will use burrows that are either dug or recently exploited by other mammals, including gray and red foxes, woodchucks, striped skunks, badgers, and opossums. Their highest density in one region of Ohio occurred in an area that was virtually devoid of tree dens but had numerous ground dens, most of which were made by wood-chucks. Their use of ground dens has also been found to be high in south-central Indiana. Interestingly, most of the ground dens they used in a south Texas study appear to have been made by armadillos. Apparently, the raccoons remodeled them by lengthening their burrows and expanding the entrances. Ground dens seem to confer some of the same temperature benefits as tree dens and may be even more protective in some areas. In one instance, an abandoned fox burrow that raccoons used was found to be both warmer in the winter and cooler in the summer than a nearby raccoon tree cavity; its average temperature fluctuation was only one-fifth that of the tree den's.

As indicated, raccoons find their winter dens in numerous other settings as well: rock crevices and caves, drains, abandoned buildings, and brush piles. Although biologists have thought that raccoons rarely use muskrat lodges, even when tree cavities are scarce, most of those in one marsh were found to den in these structures. One investigator has even discovered a raccoon with all four of its feet frozen in a muskrat lodge. The absence of suitable den trees does not appear to be a limiting factor for raccoon populations, either in marshes or on the prairies.

Following her winter sleep and subsequent mating, a pregnant raccoon typically selects a different den in which to bear her litter. As is the case for the winter den, a hollow tree is the likeliest choice for this activity. Yet again, the opportunistic mother will exploit a plethora of other sites in which to raise her litters: underground dens or burrows, especially in marshes; rock crevices; caves; abandoned mine shafts; boxes erected for nesting wood ducks; and a magpie nest have all been used.

During the day in the warmer months or when it is mild in the winter, raccoons will use their winter and litter-rearing dens for sleeping. They also rest in places offering far less protection. In marshes, swamps, and fields, sleeping sites are often simply on the ground vegetation. Though they rarely build nests, at high tide in salt marshes they have been found to construct flat platforms from salt-marsh grasses and rushes as much as 1.6 km (about 1 mile) from dry land. They also rest on tree limbs, in clumps of Spanish moss, and on the mashed-down leaf litter nests of gray and fox squirrels. One observer noted that when harassed, a raccoon on a squirrel nest first urinated, and when disturbed again, jumped out of the tree from a considerable height and hit the ground running. Though they may use some spots more frequently than others, raccoons repeatedly change their sleeping sites, even on a daily basis. In Kansas, their average occupancy period for resting dens was only 1.5 days. Winter and litter dens tend to be reused considerably more often.

The raccoon's sleeping and den sites are found throughout its home range, which is, as mentioned previously, the area where it engages in most of its activities. Den sites tend to be close to water sources. Several studies have determined that the average distance from a den to water is between 67 to 140 m (about 220 to 460 feet), with maximum lengths of 180 to 800 m (about 590 to 2,625 feet or up to about a half mile). Even those sites used for rearing young may be near the edge rather than in the interior of the mother's home range.

Although adults are reported to mostly den by themselves, in one study almost half of the females denned alone whereas only a fifth of the males did. In cold weather, raccoons often share dens, probably to conserve heat. When three to five are found together, the group often consists of an adult female with several young born that year, presumably her off-spring. An adult male and female will occasionally den together. More than two males, two females, or one of each, sometimes with juveniles, will share a den, especially in the winter. Groups of 8, 9, and 12 raccoons have been found together in one den. The largest assemblage ever found was 23, which consisted of 7 juvenile males, 7 juvenile females, 1 juvenile of unknown sex, 3 yearling or adult males, and 5 *parous* females (those that had given birth). All of them may have descended from one female.

POPULATION DENSITIES

The densities of raccoon populations vary greatly throughout their range. Undoubtedly, some of this variation is only attributable to the accuracy of the techniques used to estimate their population size. But as indicated, areas differ substantially as to their suitability for raccoons, which should greatly explain the variation in raccoon densities in different habitats. Numerous investigations of their population densities have been undertaken; only several of the more illustrative ones are considered in this chapter.

Low raccoon densities are generally those with fewer than five individuals per square kilometer, such as the estimates of 0.5 to 1.0 per km^2 and 1.5 to 3.2 per km^2 from the prairies of North Dakota and Manitoba, respectively. In the less-than-optimal raccoon habitat in the upland hardwood forests of east Tennessee's Appalachian Mountains, their densities ranged from 3.6 to 5.9 per km^2. Densities of 4.0 per km^2 during a drought and 12.3 per km^2 occurred in a subtropical area of Texas characterized by mesquite mixed with various grasses, chaparral mixed with various grasses, and live oak mixed with chaparral. (The term *chaparral* refers to an area typified by thickets of shrubby plants adapted to dry summers and moist winters.) In primarily agricultural land on New Jersey's coastal plain, there were 12.8 raccoons per km². On Quebec farmland where corn was the major crop, their densities varied from 3.6 to 21.7 per km². In eastern Virginia's tidewater region, a report indicates that there were 17.2 raccoons per km². Densities of up to 20 per km² have been noted in the bottomland and marsh areas of the midwestern and eastern United States, and even higher densities appear to be common in the southeastern states. There are accounts of 31 raccoons per km² in Mississippi and 49 per km² from an Alabama beaver swamp.

Raccoons can reach surprisingly high numbers in some areas. A strikingly high count of 68.7 per km² was found in a residential Ohio suburb. On the largely mixed dune and wetland timber sections of northwestern Indiana, their densities varied from 36 to a lofty 222 per km². Biologists Allen Twichell and Herbert Dill reported the highest raccoon density known to occur: During the winter of 1948, 100 were removed from their den trees on a 41 ha (102 acres) tract of the Swan Lake National Wildlife Refuge, a waterfowl reserve on a Missouri marsh, yielding an astounding density of about one raccoon per acre, or nearly 250 per km².

As is evident from their wide range of densities within the same area, raccoon populations often fluctuate. They may increase at a relatively rapid pace or a more gradual one. For example, an Alabama study estimated that a raccoon population in this state could, in essence, replace its numbers in 10 years, whereas a study of a Missouri raccoon population indicated that it could replace itself in 7.4 years. Again, the suitability of a particular area for the density and growth of a raccoon population varies with local conditions. Early in the twentieth century, the fabled naturalist and author Ernest Thompson Seton described their populations as "eruptive" or "irruptive," suggesting that they exhibit severe and irregular oscillations of more than 50 percent of their normal numbers. This characterization, however, was based on data from the nineteenth-century fur records of the Hudson's Bay Company. Such information may not be reliable for describing population trends because socioeconomic conditions can influence the harvesting and marketing of fur. Thus, raccoon populations may not fluctuate as much as had once been assumed.

ACTIVITY AND INACTIVITY

Several studies indicate that raccoons are *nocturnal:* They are largely active from sunset to sunrise. They generally search for food at night, with their activity peaking before midnight. Their activity rarely begins more than an hour before sunset and they might not return to their resting sites until several hours after sunrise. One investigation reports that their daily activity pattern entailed travel toward feeding areas between 4 P.M. and 8 P.M., activity in a major feeding site from 8 P.M. to midnight, activity within a minor feeding site from midnight to 3 A.M., and then a return to their resting area.

Raccoons are often active at other times. They will readily veer from their regular feeding schedule to take advantage of the availability of food or water, again illustrating their great adaptability. In coastal salt marshes, raccoons may be active at low tide and inactive at high tide, without regard for whether it is day or night. At low tide, they feast on mollusks, such as clams and oysters, as well as crustaceans, particularly crayfish, that have become exposed on mud flats, beaches, and stream bottoms. An intriguing raccoon behavior was observed near the beach at Cape Sable, Florida. During the dry season there, they dug seven wells up to 50 cm (about 20 inches) deep. They regularly visited them throughout the day to drink the fresh water that had accumulated there. The wells may have been initially excavated in their hunt for crabs or crayfish, but as fresh water repeatedly seeped into them, the raccoons apparently were prompted to revisit this area.

In an Alabama study, the raccoons were suspected to restrict their short-term movements to shifting from one small area to another within a larger, more familiar one. Portions of the larger area, especially if far from water, were rarely if ever visited. Raccoons in a part of Minnesota commonly traveled throughout their entire seasonal home range within a two-week period, though they frequented swamps, marshes, and oak woods more regularly than bogs or open fields. To visit certain food sources such as cornfields or fruit trees, raccoons will journey far from their regular home ranges. Two individuals in California regularly commuted 5.6 km (3.5 miles) each way to a plum orchard.

Raccoons that have been transplanted do not appear to be capable of *homing*, or returning to their initial location. In fact, they may wander for considerable distances. In the southern United States, raccoons have been killed or recaptured 240 km (150 miles) and 288 km (180 miles) from where they were released. However, pen-reared individuals in Michigan were likelier to remain closer to their points of release than wild ones. A considerable amount of information also exists about the dispersal of young raccoons from their birth areas. This is explored in the following chapters on mortality and reproduction.

In the winter, raccoons restrict their activity by denning for long time spans during periods of snowfall and cold weather. Their lethargic state,

or *torpor*, is not considered to be true hibernation. In genuinely hibernat-
ing mammals, such as certain ground squirrels, body temperature drops
substantially for a lengthy period during the winter. In addition, the
spring arousal from this state is usually an extended activity. During its
winter denning, a raccoon essentially stays asleep in a curled position.
However, rather than remaining lethargic throughout the winter, it becomes active on milder days. Even in areas with harsh winters, it can become wide-awake after only a slight disturbance and be ready to defend
itself. Its torpor also involves far less substantive physiological changes
than true hibernation. Yet it does undergo some increase in the size and
activity of its *islets of Langerhans*, the insulin-producing glands in the pan-
creas. Increased insulin secretion can result in below-normal blood sugar
levels, or *hypoglycemia*, which also occurs in true hibernators.

The length of their winter denning varies geographically. In the southern United States and in parts of California, raccoons are normally active
throughout the winter, except perhaps for a short time if the temperature
becomes unusually low. But where subfreezing temperatures and permanent snow cover persist, they may sleep on and off for months. In the
northern United States, their inactivity usually lasts from late November
or December until February, March, or even early April. This lethargy appears to be induced by the season's first ground-covering snowfall. After
snow has been on the ground for several months, temperature may be
more important in determining if the lethargy persists. Following one or
perhaps several days of above-freezing temperatures, they will emerge
from their dens and search for food, even in deep snow. Those with
enough fat can last for days without eating, and when it is very cold they
can obtain water by eating ice or snow. When wind velocities are high,
they tend to stay in their dens. Gusts of at least 15 miles per hour discourage their activity, especially when the temperature is $-1°C$ ($30°F$) or
lower. As suggested, raccoons can conserve heat in the winter by denning
communally.

EATING HABITS

As has been described, the procyonids demonstrate an incredibly diverse
array of feeding strategies. The kinkajou is well suited for foraging in the
trees for fruit and small animals. Ringtails and olingos are also adapted for
hunting in the trees, as are the coatis, which may also stalk their prey on
the forest floor. But of all of this family's members, the raccoon is surely
the most diverse in its feeding tactics. It appears to be equally capable of

preying on crustaceans along streams, hunting small animals, or searching for fruit in a tree.

While feeding, raccoons move around quickly for short stretches, but they will stay longer in certain areas, particularly around shallow water. They may use their jaws, but especially rely on their hands to catch prey items. In searching for food, they methodically explore holes in the ground and in trees. They hunt for crayfish and other aquatic animals by probing under rocks and in the crevices along stream banks. Yet they have also been observed searching for food in the water while staring at the sky. In marshes, they use their forepaws to dig and extract crayfish and crabs from their burrows. Small items are often eaten where they are found, whereas larger ones, even if only as big as a crayfish, are often hauled off to a bank or a boulder to be eaten. Before something is consumed, it might be visually inspected. It usually is at least appraised manually; for example, live prey may be rubbed or rolled under their forepaws before being ingested.

The raccoon typically chews its food both finely and carefully. As stated in Chapter 1, its salivary glands, which assist in the initial digestive processes, are well developed. Its digestive tract, like that of other Carnivores, is a relatively short tube from the small intestine to the anus. Food typically completes its passage through this tract in 9 to 14 hours. Their digestive tract is similar to those of other Carnivores in that it is relatively unspecialized. It lacks a *cecum*, the fermentation pouch located at the beginning of the large intestine in herbivorous mammals such as rabbits or horses.

Raccoons are capable of learning to take advantage of new foods, a behavior that may then be copied by other raccoons, including the young. The knowledge that raccoons so acquire may thereby be passed on to succeeding generations as a form of *cultural inheritance*. Examples of such transmission of learned behavior include the eating of eggs and devouring of melons after spotting broken ones.

Raccoons may join temporary feeding aggregations. Occasionally, several individuals are able to take advantage of certain foods, such as fruit or *carrion* (dead animals), which can become available in substantial amounts. When this happens, many may congregate, though they might still forage apart. In one area, nine were observed feeding concurrently along a stream in the summer, but they largely avoided being in close contact. Particularly aggressive adult males may not be included in such groups. One winter in Nebraska, nightly aggregations appeared at a feeding station. As the mating period approached, fighting incidents erupted

Raccoons may temporarily join feeding aggregations to take advantage of food sources, such as those in this pizza parlor dumpster. Photo by William G. Heatherly.

among the adults and the raccoons stopped coming to the area. The individuals in such gatherings do not form social bonds like those in other types of raccoon groups, such as mothers with their young, individuals banded together in winter dens, and certain male groups (see Chapter 9, "Social Organization").

The raccoon may well be one of the world's most omnivorous animals. They eat an astounding variety of plants and animals, and even consume human garbage. Perhaps one of their oddest meals was a platter of fried potatoes cooked in mutton fat. This epicurean delight was discovered as a solid lump in the stomach of an ill-fated individual that had broken into a Maine woodchopper's cabin. Despite it having such a diverse and, at times, bizarre appetite, a few generalizations can be made about the raccoon's food habits. First of all, the relative proportions of different types of foods vary, especially with the seasons. Another basic generalization is that in most areas they eat more plants than animals. These tendencies are a function of their opportunism, as raccoons invariably eat what is available and accessible. When food is abundant, they can afford to be more selective.

Given their broad omnivory, consummate opportunism, and widespread distribution, a complete listing of the raccoon's foods would be long, tedious, and perhaps impossible. Instead, this section will attempt to provide an informative overview of its dietary habits. The most important plant foods for raccoons appear to be fruits, berries, nuts, and seeds. In terms of the number of species consumed, their largest plant food group is the fleshy fruits. Raccoons apparently have many taste receptors that are responsive to very sweet foods. They eat both wild and cultivated fruits such as grapes, cherries, apples, peaches, plums, figs, citrus fruits, Russian olive tree fruit, and watermelons. Another fruit, that of the persimmon tree, has been part of its folklore in the southern United States, as is indicated in Chapter 11, "Raccoons and Humans." The berries that they eat include raspberries throughout their range; marlberries on Key Largo, Florida; palmetto berries in Louisiana; juneberries in Minnesota; manzanita berries in California; as well as hackberries and juniper berries, which are actually berrylike cones, in Texas. Dry berries are eaten when more acceptable foods are unavailable. Raccoons may disperse significant numbers of fruit and berry seeds, though this has not been sufficiently studied.

Raccoons also eat various nuts but especially savor the acorns of many different oak species. They have been reported to search for acorns even under 2 feet of snow. They also eat beechnuts, hickory nuts, pecans, and walnuts. Prior to their decimation from blight, American chestnuts might have helped them to store large quantities of fat for the winter. This blight was caused by a fungus that was accidentally introduced to the United States, probably at the beginning of the twentieth century.

Corn is regarded as the raccoon's most important crop food, especially when it is in its *milk stage*. This is when the kernel's internal portion is still somewhat liquid. In much of the raccoon's range, corn is heavily consumed whenever it is available. In eastern Iowa and likely other areas, corn is one of its major foods in the postwinter period. Other crops raccoons commonly eat include wheat, barley, oats, millet, sorghum, sunflowers, and the aforementioned fruits.

Raccoons invariably eat more *invertebrates*, animals without backbones, than vertebrates. The most significant animals in their diets are arthropods. This is the most numerous group of animals on Earth; it includes insects, crustaceans, and spiders. As suggested, raccoons are especially fond of crustaceans, particularly crayfish, which can be their most important animal food in some areas. In Michigan, Frederick Stuewer, one of the first biologists to conduct a comprehensive investigation of raccoon ecology, found that crayfish occurred in more than a third of the raccoon's feces or *scats*. Similarly, in northwestern Indiana, crayfish constituted more than half of the food items in their autumn scats. As indicated, raccoons living along marshes and coastal areas eat other crustaceans such as shrimp and crabs, as well as clams, oysters, mussels, sea urchins, and marine worms. In Texas, along bays near the Gulf of Mexico, the fiddler crab has been one of their major food items.

They prey on an exceedingly diverse assortment of insects, including diving beetles and dung beetles, dragonflies, grasshoppers, ants, wasps and bees, as well as miscellaneous species in their caterpillar stages. In terrestrial habitats, their noninsect invertebrate foods include snails, slugs, and spider egg cases. In eastern North Dakota, they have been known to devour great numbers of earthworms in the spring.

Throughout its range, the raccoon also regularly preys on vertebrates. Exploitation of these animals, however, can be relatively uncommon in some areas. Various reports indicate that their most commonly caught fish are the centrarchids, or those in the family of sunfish, bass (including the largemouth bass), and bluegills, and the cyprinids, or those in the carp and minnow family. Raccoons can easily catch fish in shallow water. More than a century ago, the heads of a large number of catfish, eel, pike, and perch were discovered where raccoons had been feeding in Virginia's Dismal Swamp. They have also been known to eat gars, suckers, trout, pickerel, shad, and bullheads.

Although raccoons often live around amphibians, such as frogs, toads, and salamanders, these animals typically do not constitute a large part of their diet. In North Dakota's prairie pothole region, however, tiger salamanders were found in 9 percent of the raccoon scats. Reptiles also do not appear to be significant in their diet. Raccoons occasionally eat snakes, such as water and garter snakes, as well as various lizards. In some southeastern marshes, they also dine on alligator eggs. Various freshwater turtles, including cooters, sliders, mud turtles, snapping turtles, and softshell turtles, as well as their eggs are also part of their diet.

In several areas, raccoon predation on sea turtle eggs is so common that it has become a concern. On U.S. beaches, raccoons may prey on between 16 to 87 percent of the sea turtle nests in an area. Obviously their impact is highly variable; it depends on such factors as the relative numbers of turtles to raccoons. Sea turtles are protected in the United States by the Endangered Species and Marine Turtle Recovery acts; hence, the beaches where they nest must be managed for their recovery. On Florida's Cape Canaveral National Seashore, most of the turtle nests belong to a threatened species, the loggerhead turtle, though a small number of endangered species such as the green turtle, the leatherback, and the hawksbill also nest there. In one study, putting protective screens over their

Young raccoon digging up a common snapping turtle nest in a sand pit near Quabbin in central Massachusetts. Photo by Thomas J. Maier, courtesy of Thomas J. Maier and the United States Forest Service.

nests was found to be more effective in safeguarding eggs than either re-moving about half of the raccoons from the area or spiking the eggs with vomit-inducing estrogen.

Though raccoon predation on birds is also common, it is highly vari-able in its intensity and the types of birds taken. As with their other foods, raccoon exploitation of birds is a function of their availability and the op-portunities for pursuing them. At times, their impact on bird populations can be devastating, especially when they are introduced into an area where they or other predators do not occur. On coastal islands off of Mas-sachusetts, the release of raccoons and foxes almost entirely eliminated the production of young in herring gull colonies and the abandonment of some colony sites. Raccoons also have the capacity to inflict severe losses on burrow-nesting seabirds. Probably introduced onto British Co-lumbia's Queen Charlotte Islands in the early 1940s to strengthen its fur industry, raccoons are now widespread there. On one of these islands, as few as three or four individuals were implicated in the deaths of dozens and possibly hundreds of ancient murrelets. Apparently, they killed many more of these birds than they needed for food. Such surplus killing is not

Raccoon digging up a mallard duck's nest in Manitoba, Canada. Their predation on such nests can seriously impact waterfowl populations. Photo by Glenn D. Chambers.

unusual among various mammalian predators. Because these islands contain most of the world's ancient murrelets, the raccoon appears to be a significant threat to their survival.

In Raymond Greenwood's study of the raccoon's food habits in North Dakota's prairie pothole region, birds and their eggs occurred in 34 and 29 percent of their scats, respectively. The remains were primarily from red-winged and yellow-headed blackbirds, common grackles, and brown-headed cowbirds. They also killed ducks, marsh birds known as rails, and coots, birds that resemble ducks. The coots accounted for more than half of the large birds they killed, though more than half of the eggs that were taken from large birds were from waterfowl. Raccoons also prey on various songbirds, woodpeckers, pheasants, and quail (mostly bob-whites), and will snatch the eggs of pheasants, quail, and turkeys. Some of their more unusual avian victims are short-eared owls, common moorhens, great blue herons, black-crowned night herons, and double-crested cormorants. One bold individual was observed robbing a red-tailed hawk's nest about 15 m (about 49 feet) above the ground.

Raccoons are well known as predators of waterfowl, particularly ducks and geese, and their eggs. Their predation on eggs and hatchlings can considerably impact duck populations. Undoubtedly, their effectiveness as waterfowl nest predators is enhanced when the water in marshes is low and their quarry thus becomes more accessible. They also often feast on ducks and geese that are crippled during the hunting season. Within the last several years, their predation on birds at the Bear River Migratory Bird Refuge in northern Utah has become so severe that refuge managers have instituted control measures such as the destruction of raccoon den habitat.

Raccoons also eat numerous mammals, mostly small rodents such as meadow voles and various mice. In a Michigan study, voles constituted between a quarter to a third of their spring diet. Larger rodents, including gophers, ground and tree squirrels, and muskrats are also taken. They will seize muskrats from traps and feed on their carcasses. In the spring, they often prey on young muskrats or *kits*. In one Wisconsin marsh, these kits and crayfish were their major summer foods. Raccoons also prey on shrews, moles, rabbits, and occasionally jackrabbits. They will prey on other Carnivores, such as mink, though probably not very often. They also eat the carrion of wild and domestic large mammals such as deer, cows, and horses. Finally, raccoons may resort to cannibalism, with larger individuals feeding on smaller ones.

Though they eat a remarkable variety of foods in so many different environments, raccoons often follow similar patterns of seasonal diet change. In most areas, they only consume more animal than plant foods in the spring. As previously noted, crayfish tend to be their principal animal food at this time, followed by insects and small vertebrates. Corn is frequently a major component of their spring diet, and acorns can be especially valuable in early spring before other foods become available. In North Dakota's prairie pothole area, their use of plants decreased from April through July as they increasingly depended upon animals. Insect feeding climbed throughout the spring, as did their predation on birds and their eggs, which was augmented by the arrival and nesting of migrant species. Their consumption of other animal groups in that region, however, was not consistent with their pattern of spring reliance on insects and birds. The percentage of mammals in their diet decreased from 30 percent in April to only 8 percent in July, while amphibians and crustaceans were mostly consumed in July.

In the summer, raccoons largely eat fruit. In a Michigan study, fruit composed more than 77 percent of their summer diet. As in the spring, corn is also a significant food in many areas during this season. Crayfish, followed by insects and small vertebrates, are often their most common summer animal foods.

In the autumn, plants remain their most important food in many regions. Corn and fruit such as grapes are eaten in abundance. Animals, especially waterfowl that have been injured by hunters, tend to be more vital to raccoons that live around marshes. In a study of raccoons in Illinois's river bottomlands, between 1,000 to 2,000 crippled geese made up the majority of their autumn diet. Acorns are frequently their most common winter food, supplemented by any remaining corn and fruit. In some areas, they exploit various invertebrates and small vertebrates in the winter. Trapped muskrats and crippled waterfowl often continue to support marsh raccoons at this time.

Although this information on seasonal food habits depicts some general patterns, it cannot be overstated that the raccoon's feeding habits are strongly influenced by what is available. For example, during one summer in southwestern Washington State, they mostly consumed the animals living along river mud flats rather than plants, which are usually heavily utilized at that time. In addition, raccoons in a part of Maryland were found to rely considerably on insects late one winter. As mentioned, in most areas, their greatest intake of animals appears to usually occur in the spring.

PATTERNS OF WEIGHT CHANGE

During the winter in the northern parts of their range, adult raccoon weights can drop substantially, a pattern that surely occurs in many other animals. In Minnesota, a 50 percent decline in adult weights occurred from late November through mid-March. These weight dips are less severe and occur several weeks later in the southern states. Yet even in these areas, raccoons may lose between 14 to 28 percent of their body weight at this time. Individuals likely put on extra weight while food is readily available to prepare for the upcoming harsher conditions. The yearlings' early autumn weight gain in Minnesota has been described as spectacular.

In the winter, the raccoon's fat stores become depleted for several reasons. First of all, staying warm at this time simply has a higher metabolic cost. Their food resources and therefore food intake may also be reduced. In addition, because snow covers much of their edible vegetation, they may have to rely more on predation, which is energetically expensive. Northern areas that have colder temperatures and vegetation made inaccessible by deep snow must be the most energetically demanding places for raccoons in the winter.

Gehrt and Fritzell suggested that the raccoon's sudden weight loss in Texas in March might be partially attributable to the energetic costs of their mating behavior. Furthermore, Texas summers are long, hot, and humid. The weight loss of raccoons in this and similar environments during this time could be a reaction to the heat and a decreased need for the energy provided by subcutaneous fat. Indeed, the raccoon's capacity for thriving in a variety of environments has been attributed to several of its physiological qualities: its well-defined cyclical patterns of both fat content and thermal conductance, its high capacities for evaporative cooling and heat tolerance, and its high metabolic rate.

In spite of its remarkable success in so many locales under such varied conditions, the raccoon has obstacles to its survival as does any animal. The factors that constrain its survival and that may ultimately lead to its demise are examined in the next chapter.

7

Mortality and Disease

P eople are without question the raccoon's major predators. *Harvesting* (hunting and trapping by humans) and collisions with motor vehicles are frequently the most significant causes of raccoon mortality. In one Iowa study, for example, harvesting by humans accounted for 78 percent and vehicle accidents for 10 percent of all of the determinable raccoon deaths. The relative importance of other mortality factors apparently varies with the degree of exploitation by humans. Thus, in a Minnesota population that was only lightly hunted, more than half of the deaths were caused by other factors: starvation, extreme parasitism, and disease. The effects of all these factors on the raccoon's survival are discussed in this chapter.

PREDATORS

Predators such as bobcats, coyotes, and several owl species such as the great horned owl will occasionally kill raccoons. Others that generally have much less of an impact on their populations include cougars, wolves, fishers, red foxes, and gray foxes. With the possible exception of the first two, these mammals were probably never significant predators of raccoons, and only circumstantial evidence points to the wolf having had an impact on their distribution (see Chapter 5). Furthermore, except for the foxes, they are now rare or absent throughout much of the raccoon's range. Alligators also prey on raccoons, probably mostly on juveniles. In the few cases where it has been studied, raccoons have only accounted for a small fraction of the alligator's diet. However, on Georgia's Cumberland Island, raccoons were found to be the alligator's most common food item.

MALNUTRITION

Malnutrition and its effects are other causes of raccoon mortality, especially in the late winter and early spring. The young are particularly vulnerable to starvation. Given their higher surface-to-volume ratios, younger raccoons have relatively more exposed surface area than the adults. Thus, they lose proportionately more heat to the environment in cold weather. Juveniles also have less body fat than adults, both in total and as a proportion of their size, and therefore fare worse when it is cold. In the aforementioned Minnesota population, starvation, mainly among yearlings emerging from their winter dormancy in March, was the major cause of mortality. Another example of the effects of low food availability comes from a Maryland population: These raccoons lost up to 50 percent of their weight when production of the *mast* crop, that of acorns and other tree foods, failed for two successive years.

DISEASE AND PARASITISM

Being in a weakened state from malnutrition can result in a higher susceptibility to other mortality factors such as disease and parasitism. Raccoons are periodically affected by several diseases, including respiratory ailments such as pneumonia and chronic pleurisy, an inflammation of the membrane around the lungs. Yet it appears that the only diseases that can seriously impact their numbers are canine distemper and rabies. Such *epizootics*, diseases that affect many individuals, can spread rapidly in raccoons, both within localized populations as well as over wide areas. Most of their other diseases are more confined.

Distemper, which is caused by a virus, can be a significant factor in the decline of both rural and suburban raccoon populations. Periodic outbreaks of this disease have been reported throughout the United States. In 1968, raccoons of all ages succumbed during an outbreak in Clifton, Ohio. Individuals were found dead or dying in the woods as well as on porches. Within about a year, the population plummeted from 145 to 51 animals. These eruptions can be so devastating that by the early 1990s,

the city of Scarborough in Ontario, Canada, began to vaccinate its raccoons against distemper. This program reduced its prevalence and apparently did not precipitate the population boom that some of its critics
had predicted. Another viral disease, encephalitis, which causes inflammation of the brain, may follow distemper. Various wild Carnivores suffer from a *distemper complex* that includes both distemper and encephalitis,
and can mimic the effects of rabies. Therefore, a raccoon that is stagger-
ing about might be encephalitic rather than rabid.

Rabies is the other disease that can significantly impact raccoon populations. It is caused by a *neurotropic* virus, which means that it grows on the
host's nervous tissue. It is carried in the rabid animal's saliva. Following
transmission by a bite, the virus attaches to the nerves and grows until it
reaches the brain where it causes an inflammation. Its most obvious symp-
tom is a change in behavior. Friendly animals may become belligerent and
timid ones can turn bold. Contrary to popular belief, rabid individuals
may be passive.

Concern about rabies in raccoons has been prevalent for almost 40
years. Between the 1950s and 1977, rabies began to spread northward,
mainly within Florida and Georgia. Then, in the late 1970s and early
1980s, what has been described as an epidemic began to sweep through
the Middle Atlantic states. Undoubtedly, the movement of several thousand Florida raccoons to Virginia and North Carolina to restock their
populations—shipments that are now known to have included rabid individuals—facilitated the spread of this disease. Through the mid-1990s,
rabies outbreaks were reported in various northern Middle Atlantic and
New England states. In 1992, more than 4,300 rabid raccoons were documented in the United States, a much higher number than had been reported for any other animal. Louisiana appears to be one of the few southeastern states without any incidence of this disease, though its state
biologists are still concerned about its possible occurrence.

In recent years, predictions indicate rabies could reduce raccoon populations by substantial levels in much of the northeastern United States.
People and their pets have a potentially high exposure to this disease
given the number of raccoons that live close to urban human populations.
By now, efforts to control rabies in raccoons have prompted the imple-
mentation of vaccination programs as far north as Toronto. Even though
the disease's major *vectors*, or carriers of the disease, in this area are red
foxes and striped skunks, rabies may be the primary threat to raccoon
populations in this and other urban locales. The Toronto population increased by 40 percent within three years of the plan's initiation, despite
the contention that widespread raccoon vaccination programs are unfea-

sible. Of course, those who regard them as pests may not appreciate such efforts.

Although rabies does not spread readily from raccoons to other species, raccoons can be reservoirs for diseases that affect humans as well as livestock. In the southeastern United States alone, they carry more than a dozen pathogens that can cause illnesses in other species, including humans. There seems to be relatively greater concern about raccoons serving as reservoirs for certain diseases—leptospirosis, tularemia, Chagas' disease, and rabies—than for others that they carry, such as tuberculosis, listeriosis, and the fungal disease histoplasmosis. As mentioned, they can carry encephalitis, including such forms as St. Louis encephalitis, fox encephalitis, as well as both Venezuelan and eastern equine encephalomyelitis.

Leptospirosis is a bacterial disease caused by a microorganism, *Leptospira* sp., that affects the kidneys and liver. Tularemia, known as rabbit fever, is another bacterial affliction; it is caused by the microorganism *Francisella tularensis.* In the southeastern states, these two diseases have their largest reservoirs in raccoon populations. Both are transmissible by direct contact with raccoons or through water contaminated with their urine or feces. In some areas, the high incidence of their leptospirosis is regarded as a public health problem. Listeriosis, yet another bacterial disease, affects the central nervous system and causes toxins to be released into the blood. In this *septicemic* or blood-poisoning form, a raccoon can suffer from tiny liver lesions and a swollen spleen. Chagas' disease is also known as trypanosomiasis; it is caused by a *protozoan*, a one-celled animal.

Heavy loads of internal and external parasites may exacerbate maladies such as distemper and rabies in raccoons. Intriguingly, however, well-fed individuals can carry large numbers of certain parasites without apparent ill effects. Dozens of internal parasite species occur in raccoons, especially worms such as roundworms, tapeworms, flukes, and spiny-headed worms. They commonly ingest parasite eggs while feeding on animals or feces. Parasitism in the raccoon has been so well studied that several roundworm parasites are named after it. One, *Gnathostoma procyonis*, can cause widespread lesions in their stomach and other tissues. Another, *Baylisascaris procyonis*, which seems to be especially common in raccoons in the northeastern and midwestern states, could pose a serious health threat to children who accidentally ingest raccoon feces filled with its eggs. Raccoons also may suffer from a lungworm, *Crenosoma goblei*, which is a major cause of chronic respiratory infections, especially in malnourished individuals. Large infestations of some parasites can interfere with bodily functions, such as by blocking the opening of the small intestine.

Raccoons also harbor a wide variety of external parasites, including ticks, fleas, and lice. Bites from one tick species, *Dermacentor variabilis*, as well as from lice, can transmit tularemia, which as mentioned is also spread in other ways. Certain mites can cause a skin disease called scabies or sarcoptic mange, which causes hair loss. Ear mites are occasionally a problem for them as well.

Nevertheless, various diseases, including those associated with parasitism, may only substantially regulate raccoon populations when they are at very high densities. As has been found in other animals, raccoons may suffer from physiological stress at such densities, which could make them more vulnerable to the effects of their parasites. Concentrations of internal parasites might decrease in the winter in some areas as raccoons curtail their food intake while dormant. The parasites may then die off or become sterile.

MORTALITY RATES: COMPARING THE SEXES AND AGE GROUPS

Several recent long-term studies have provided important details on the nature of survival and mortality in raccoon populations. In general, this research indicates that survival patterns vary substantially among the sexes, within and between years, and across different age groups. Various researchers have suggested that males and females differ in their vulnerability to hunting and trapping. An Iowa study indicates that a lower percentage of adult males than adult females survived harvest activities. Raccoons tend to have a *polygynous* mating system in which a male mates with more than one female (the details of which are discussed in Chapter 9). In such mating arrangements, males generally intensify their movements and may broaden their home ranges to enhance mating opportunities during the breeding season. This heightened activity could increase their susceptibility to getting killed during a late winter harvest. Even if no harvest occurs, the males' increased movements should at least burden them with additional energy costs and magnify their exposure to predators, disease, and accidents.

Recently, however, Michael Chamberlain and his research team found the opposite trend in a raccoon population in central Mississippi. From 1991 to 1997, the average annual survival rate of the adult males (63 percent) was significantly higher than that of the adult females (50 percent). (In this context and analogous ones herein, the term *significant* means that statistically, less than a 5 percent probability exists that the indicated difference between the compared values is due to chance.) However, the circumstances in this area are complicated, with survival rates often shifting both throughout the year and from one year to the next. To fully assess this variation in survival, these investigators defined three intervals: the breeding-gestation period (February 1–May 31), the *parturition* (birth)–young rearing period (June 1–September 31), and the fall-winter period (October 1–January 31).

They found that no significant differences existed between male and female survival rates during the breeding-gestation and fall-winter periods. But when the females were caring intensively for their young during the parturition–young rearing period, they survived at a higher rate (93 percent) than did the males (83 percent). Female survival, however, was considerably lower during the breeding-gestation period (65 percent) than it was during the parturition–young rearing period. The males had a significantly higher survival rate (91 percent) during the fall-winter than they did in the other periods. Finally, the males' survival rates differed between the years whereas that of the females did not.

Interestingly, canine distemper was most widespread in this population during the breeding-gestation period. As suggested, the males' increased activity at this time may result in their having greater exposure to diseases. Yet, these researchers presented other perspectives on the prevalence of distemper during this period. First, the increase in male-female interactions at this time should cause both sexes to have a greater susceptibility to catching and spreading such diseases. In addition, although the males may increase contact with one another during this period through confrontations, they might also avoid each other by scent marking and other behaviors while maintaining territories and seeking mates.

The specific causes for the various mortality rates did not differ between the males and females in this area. Of the 69 deaths investigated, 22 (32 percent) were legally killed by hunters while none were taken by trappers, 5 (7 percent) were attributed to other human factors, and 11 (16 percent) died from natural causes. The reasons for the deaths of the other raccoons, almost half of those examined, could not be determined. Eight of these were found near individuals that were dying from distemper, so they may have also died from this disease. Even though legal harvests accounted for most of the deaths for which causes were known, the basis for mortality in a population is difficult to understand when the causes of so many of the deaths in a sample cannot be determined.

Gehrt and Fritzell's recent study on south Texas's Welder Wildlife Refuge has also yielded valuable information on raccoon mortality. The refuge's raccoon population has been shielded from harvesting since the early 1950s. There was no significant difference in the survival of its male and female adults; the survival rates for each were greater than 80 percent. (This differs from the pattern in the Mississippi study in which the females had lower annual survival than the males.) Of the 18 Texas raccoons radio-marked as adults, 78 percent remained alive after one year, 50 percent after two years, and 44 percent after three years. Apparently, the individuals on this refuge had a fairly good chance of surviving once they reached adulthood. Of a total of 74 radio-marked individuals, 23 had died and the fates of 7 were unknown. The deaths were attributed to disease (eight cases), predation (five), poaching (three), vehicular accidents (one), research-related incidents (two), emaciation or injury (one), and unknown factors (three). The disease seemed to be distemper but this was not confirmed. Even though numerous predators were in the area, they seemed to have little effect on the adults. Coyotes killed just two, and alligators killed only three raccoons despite these latter two species staying near one another around artesian springs during a drought.

The raccoons in this study were placed into three age groups: juveniles (less than 10 months of age), yearlings (10 to 22 months), and adults (greater than 22 months). In other studies, the same categories may be composed of different age intervals. For example, a yearling could also be classified as one between 13 to 24 months old. The survival of the juveniles was examined at four times during their growth: the nestling period, from birth to emergence from the *natal* or birth den; the mobile period, from emergence to September; the autumn period, September to November; and the winter period, December to February. Of the 33 litters that were observed, 18 suffered either partial or total mortality during the nestling period. Depending on the estimation procedure that was used, only between 52 to 65 percent of the nestlings survived to the time of emergence from their dens. Thus, close to half of them may have perished. Their fates, however, improved dramatically as they developed. The survival rates of juveniles in the older age brackets were relatively high, with values ranging from 88 to 100 percent.

Of those that were identifiable, 16 of the birth dens were in the ground or in burrows, 10 were in brush piles created by brush clearing, and only 2 were in tree cavities. The litters from the brush piles were relatively successful (90 percent) at surviving to emergence compared to those in the other den types. Brush piles also seemed to confer better protection from predators than ground dens, which probably explains why multiple litters were reared in them. Though juveniles occasionally disappeared from the ground dens, why or how this occurred was not known. Just one of the two tree cavity litters was successful; with so few tree cavities in this area, the mothers apparently were compelled to utilize other sites. Unfortunately, very few studies of nestling survival have been conducted, and thus determining if the patterns in the Texas population are representative of those in other areas is impossible.

In an Illinois study in which the survival of very young raccoons was considered, more male than female young were present in a group of younger than two-month-old raccoons than was found among the embryos. Though this was not a statistically significant difference, it suggests that the females had a higher mortality rate during the first two months of life, assuming that there were equivalent numbers of newborn males and females. Studies of survival probably miss important incidents if the monitoring does not start as soon as the young are big enough to wear radiocollars or other tracking devices. To obtain the fullest possible understanding of survival patterns, monitoring of the juveniles should begin at birth.

The closest comparable study to the one in Texas may be from Iowa, where the survival rate for juveniles from one to five months old was reported to be 65 percent. Nestling survival may have also been high in Iowa at this time as no mortality of the young was detected before they moved from their natal den sites. Most of the natal dens in this area were in tree cavities, which typically afford good protection. After raccoons emerge from these birth dens, their survival varies. In Alabama, relatively little juvenile mortality occurred during the summer, which parallels the findings for the high survival of older juveniles in Texas. Not surprisingly, survival was found to be low for juveniles exposed to an intense harvest in Iowa.

In the Texas study, all of the male raccoons apparently had either died or dispersed from their birth area by the end of their yearling period. The females, on the other hand, were *philopatric*, meaning they remained in their birth area. (The implications of this phenomenon for the raccoon's mating biology are addressed in Chapter 9.) Only 2 of the 17 females that were either born there or radio-collared as juveniles died when they were yearlings. Raccoon survival during dispersal has not been well studied, but based upon research on other mammals, it would not be expected to be high. In Texas, the survival rate of the yearlings was lower than that of the adults, but this difference was not significant. Similarly, the annual survival of the yearling and adult females did not appear to differ in an Iowa population. Studies from other northern populations in both Minnesota and North Dakota, however, alternately showed that the yearlings in these places had lower survival.

Finally, an average longevity of just 1.8 years has been estimated for raccoons living in places as far apart as Missouri and Manitoba, Canada.

Yet in a part of Alabama with comparatively mild winters and relatively little hunting and trapping, an average longevity of 3.1 years was calculated. Researchers believe that the majority of raccoons live less than five years, and that only about one out of a hundred lives to be seven years old. So though it is conceivable for a wild raccoon to live for as much as 16 years, most die at a young age. The disparity between their potential and actual longevity is thus largely attributable to the high number of juvenile deaths. The juvenile mortality rate in the Manitoba study, for example, was estimated to be greater than 60 percent.

Because this chapter focuses on death and disease, it may come across as somewhat grim. The next chapter, however, reviews the raccoon's reproductive biology and the development of its young, and therefore is more uplifting.

8

Reproduction
and
Development

Various elements distinguish the raccoon's reproductive biology, such as its breeding period, pregnancy rates, and litter size variation, as well as when it attains sexual maturity. To understand how its young develop, one must also consider how the raccoon mother prepares for the birth of her cubs and the newborns' characteristics, nursing behavior, growth patterns, and ultimately, their strides towards independence. Together, an appreciation of the raccoon's reproductive biology and its development continues to enhance our understanding of its success.

REPRODUCTION

Throughout much of North America, raccoons mate from January through March, with a peak of activity in February. Their pregnancy lasts for 63 to 65 days, though extremes of 54 and 70 days have been reported. Thus, most of the litters are born in April. In an Illinois study, for example, their average birth date was April 18. As might be expected, however, their breeding periods vary considerably. Raccoons may mate over a more extended time span, such as those in Kansas that mate from December to June, even though their breeding peak is also in February. In many parts of the southeastern United States, mating occurs later and persists longer than it does in the central states. Along parts of the Georgia-Florida border, mating lasts from February until August. The breeding in this area peaks relatively late as well, with the largest percentage of females mating in March. Comparable breeding peaks have been observed in South Carolina, Louisiana, and Texas. Interestingly, raccoons may also delay mating around their northern range limits in Minnesota, North Dakota, and Manitoba. In these areas, their mating season is from February through June, with a peak in March. Most of this region's litters are therefore born in May, with some arriving as late as September. Hence, northern cubs may have average birth dates that occur later than those in certain lower latitudes.

Photoperiod, the amount of daylight, appears to play the most critical role in the initiation of raccoon reproduction. In one experiment, both males and females that were exposed to artificially increased day length became ready to breed two to four months earlier than those not exposed. In mammals, the series of changes in reproductive readiness that occur in sexually mature females is called the *estrous cycle*. It is a sequence of physiological and structural alterations that normally culminates in a period of heat or sexual receptivity known as *estrus*. In the previously mentioned study, the initiation of the raccoon's estrous cycle was not as affected by either temperature or the amount of snow as it was by photoperiod. Nevertheless, late snows and low temperatures can limit their movements, and thus mating can be delayed in raccoons that are capable of breeding. This apparently explains why those living in far northern latitudes, where heavy snows and cold temperatures prevail, may breed later than those in the southern United States. In any event, raccoon mating usually occurs immediately after the start of their spring activity, though another period of inactivity comparable to their winter lethargy may follow it.

Males may not be sexually active throughout the year. Differences in testis weight can be substantial during the year, suggesting that this organ's weight may be a useful indicator of a male's reproductive condition. When at their largest, the testes from an Illinois raccoon weigh about 2.8 times more than when they are at their smallest; they are at a minimal weight during the summer and grow to maximal size in December. These weights can also continue to increase throughout the winter, as has been found in raccoons in North Carolina. Yet testis weight alone may not fully reveal a male's breeding status. Though some males' testes may be devoid of sperm for three- to four-month periods, it can be found year-round in the *epididymides*, the tubes that emanate from the testes.

In an Illinois study, Glen Sanderson and A. V. Nalbandov determined that a newborn's testes grew at a uniform rate from birth until about 10 months of age. Males reach sexual maturity and thus may be physiologically capable of breeding when they are about a year old, but they typically do not breed until their second year. Ordinarily, they simply do not get the chance to mate because they come into breeding condition about three to four months after the adult males. By then, most of the females are already pregnant. However, these yearlings could father some of the litters that are born toward the end of the breeding season.

The ovaries of the adult females also follow a pattern of seasonal weight change. In some areas, ovarian weights are at their lowest in July. Then their weight slowly increases until November, declines until December, and then increases again until their annual peak in April when it again begins to decline. At their peak, the ovaries are only slightly heavier than they were in November. Such weight-change patterns, as for those of the male's reproductive organs, can vary according to the winter's severity. In the Illinois study, the maximum ovarian weights were on average about 1.6 times heavier than their minimum July values. Along with this weight gain, significant changes take place in preparation for pregnancy: the uterine horns become more convoluted or twisted, their walls thicken, and they often become more *vascularized*, which means that their blood vessels proliferate. The greater number of blood vessels in this area brings more oxygen and nutrients to the embryos.

The Illinois juveniles' ovaries were found to grow at a fairly steady rate from the cubs' birth in the spring through November. They reached their maximum weight at this time again, some three months before the peak of the mating season. Their ovaries typically declined in weight through January, a trend that could last until March. Nevertheless, a large proportion of the yearlings successfully bear litters in many areas. In west-central Illinois, 73 percent of the females successfully mated before their first birthday, and between 59 to 73 percent in the northern part of the state did so. A wider range of yearlings, 38 to 77 percent, were found to successfully breed in a Missouri study. In some populations, the percentage of yearling females that breed may be substantially lower.

Clearly, the pregnancy rates of yearlings are typically much lower than those of the adults. William Clark and his colleagues conducted a study in Iowa that revealed a 59 percent average pregnancy rate for yearlings, whereas that of the adults was 91 percent. Adult pregnancy rates commonly may be more than 90 percent in some areas. Nutrition appears to play an important role in raccoon productivity. In one experiment, increasing the protein levels by 40 to 50 percent in the diet of captives resulted in a 23 percent increase in their pregnancy rate. Similarly, raccoon pregnancy rates and litter sizes appear to be higher in South Carolina in years when the acorn production is greater.

Yearling pregnancy rates have far more annual variation, and this may be a highly influential factor in the raccoon's population dynamics. However, the available data is likely insufficient to demonstrate a significant relationship between yearling pregnancy rate variation and raccoon population changes. Although several studies suggest that yearlings and adult females breed at about the same time, sexual development in yearling females may be so delayed that, like the yearling males, they do not mate until after the adults do. This delay in mating, as well as their lack of experience and maturity, could account for their lower and more variable pregnancy rates.

Evidence for raccoons producing more than one successful litter per year in the wild seems to be lacking. Yet both adults and yearlings that fail to become pregnant during their first spring estrus or those that lose a litter shortly after birth may experience a second estrous cycle about two to four months later. Such females, if fertilized, might be able to produce an unusually late litter. And in one study, the females started new consortships with males just a week or two after losing their litters. Yearlings that do not become pregnant during their first cycle might not breed until the following year, perhaps as a result of their delayed sexual development.

In one experiment, some of the females that were induced into having an early estrus were able to produce a second litter just after weaning their first. However, a third estrus apparently will not take place in those that lose a second litter. Surely, late births occur in many areas, but these have little chance of surviving where the winters are harsh. It is highly unlikely for raccoons in such regions to even produce second litters. And if they did, the young from these litters may be smaller than those from the previous litters and thus less fit to deal with the rigors of winter. Clearly, more data are needed on the survival of late-born young from both first and any possible second litters.

In west-central Illinois, more than 45 percent of the females that were greater than two years old and almost a third of the 73 percent of the yearlings that had mated had two sets of *placental scars* during the breeding season. These scars are dark bands that form when the newborn's placenta tears away from the inner lining of the uterus. As is discussed later, the number of these can be used to determine the size of the raccoon's litter. Having two such sets of scars from the same season revealed that these individuals had been pregnant twice during this time. Yet, as just mentioned, a female rarely becomes pregnant twice in one year, especially within her first breeding season. This seemingly anomalous finding could be the result of many of them losing their first litters by resorption of early embryos, abortion of fetuses later during pregnancy, or the death of their newborn cubs. After any of these events, a female might still be able to successfully mate. Though these findings suggest that the production of young from second litters could be significant, again, the survival of raccoons born toward the end of the breeding season is doubtful.

As a female nears estrus, the *vulva*, the area around her genitals, starts to swell and becomes redder. She becomes receptive to copulation a week to two after these changes begin, but only for about three to six days. The vulva, though, does not resume its normal appearance for another three to four weeks. In raccoons that do not become pregnant, the interval between ovulations is about 80 to 140 days. Raccoon copulation has rarely been observed. Apparently, it can last for an hour or more. The male's *baculum* or penis bone is acutely curved at its far end, which makes it easy for him to maintain penetration. One report described how a male would alternately thrust and pause, resting his head along his partner's back. Although the female was passive during this activity, after about a half-hour she clearly indicated that the session was over: She turned her head, laid back her ears, and snarled at her suitor.

It appears that the raccoon is an *induced* ovulator. In this condition, the stimulus of copulation is needed for a female to shed the eggs from her ovary. Nevertheless, females that have been isolated from males have also been known to undergo *spontaneous* ovulation, for which such stimulation is unnecessary. The opposite situation has also been observed. Laboratory rats that are normally spontaneous ovulators can be provoked to ovulate by stimulation of the cervix. Apparently, induced and spontaneous ovulation can be regarded as phenomena at opposite ends of a continuum. This should explain these seemingly contradictory observations.

Raccoons may be induced ovulators as a consequence of their having *semidetached* sex lives. Because males and females normally do not cohabitate nor live within cooperative groups, it could be to the female's advantage to wait for the best available male to induce her to ovulate. If she were to ovulate spontaneously and no males were nearby, she might miss an opportunity to have her eggs fertilized or at least forego the chance of being impregnated by a superior male. Other induced ovulators that exhibit such social arrangements include the cougar and various weasels. Induced ovulators are suspected to be able to extend their period of heat if they have not mated, which would also be advantageous in species with loosely organized social structures in which the availability of a mate is not predictable. On the other hand, spontaneous ovulators, which include humans, shed their eggs whether they mate or not. Among the Carnivores, such diverse species as the gray fox and spotted hyena are spontaneous ovulators. These species have mating systems in which co-habitation occurs, and the males and females may share a den. In these circumstances, a female should ordinarily have less to risk than an induced ovulator if she were to miss an opportunity to mate.

Raccoons typically have eight mammae, though six have occasionally been reported. Newborn sex ratios are usually about 1:1 (one male to one female). As mentioned, litter size can be estimated by counting the placental scars on the mother. Care must be taken in counting them, as sets from more than one litter may be present; to find sets from more than two litters in an individual is uncommon. The ones from the most recent litters are normally much darker than those from the older litters. Reproductive data on raccoons, including that on litter size, is often gathered from the carcasses at fur houses, the establishments where trappers and hunters sell their quarry. At these sites, the pelts are usually cleaned and then sold to furriers. Researchers studying raccoons on a large scale, such as the biologists who work for wildlife agencies, as well as those collecting data on a more local level regularly obtain information from these places.

Their litter sizes range from one to eight young, with averages normally varying from two to five depending upon the location. The raccoons in one North Dakota population had a relatively high average litter size of 4.8. The average adult litter size in both Illinois and Missouri studies was 3.6. The statewide average of the adults in Virginia was 3.4, which is rather high for raccoons in the southern United States. In studies from more southern states, the average litter sizes were all smaller and similar to one another: Tennessee (2.6 and 2.8), Alabama (2.5), South Carolina (2.8), and North Carolina (2.7 and 2.9). Large litters may be more likely to occur in the northern part of the raccoon's range to compensate for the presumably higher mortality of the young. Having more offspring, as well as delaying sexual maturity, might enhance their fitness in those locales. Yet despite these seeming trends, no clear-cut patterns of litter size variation in the raccoon are evident.

Several studies have demonstrated that yearlings also tend to have smaller litters than adults. In the Midwest, the litters of yearlings are typically 10 percent smaller than those of the adults. This disparity can be even more substantial: An Iowa study reported that the average litter size of the yearlings was 3.1 whereas that of the adults was 3.8. Even though they often have smaller litters, and as stated, lower and more variable pregnancy rates than adults, the yearling females' reproductive contribution can be significant, in part due to their abundance. In Missouri and Illinois studies, they accounted for 46 percent and 49 percent, respectively, of the breeding populations, generating 31 percent and 39 percent of the annual reproductive output in these areas. In Iowa, however, their contribution was much lower. These yearling females constituted an average of 31 percent of the breeding population and produced only 19 percent of the newborns.

DEVELOPMENT

For several days prior to giving birth, the mother-to-be remains in her den. During this period, she becomes more aggressive toward other raccoons. Although she will often chew and scratch the wood in the cavity, creating a primitive bed for the newborns, she does not build a nest. As each cub is born, she frees it from its embryonic membranes and licks it dry. She eats these membranes and the placenta, assumedly for their nutrient value and perhaps to keep the den tidy. At birth, a newborn weighs about 60 to 75 g (2.1 to 2.6 ounces). Its crown-to-rump length is approximately 95 mm (about 4 inches). The eyes are closed; they usually open within 21 days, though this may take up to a month. Its tiny ears are black and hairless, and the ear canals are sealed.

At birth, a raccoon is mostly covered with hair, though its undersides are almost bare. Their fur may be black, dark gray, dark brown, light tan, or even yellowish. Their skin is a dark maroon gray and their hand and foot pads are a deep maroon to black. Though the hairs on the back are sparse, the area becomes well covered by the end of the first week. The outer or guard hairs appear at six weeks and the adultlike pelage is complete at seven weeks when their first molt occurs. At first, the tail rings may only be indicated by sparsely furred, pigmented skin. They may not become evident at all until the cub is about a week old, and might not be fully furred for another two weeks. Short black hairs cover the darkly pigmented muzzle, foreshadowing the mask that becomes fully furred in about two weeks. The newborn raccoon has soft facial vibrissae, or whiskers.

The newborn is relatively small and helpless, and its head is disproportionally large for its body. This is typical of *altricial* mammals, those that are relatively immature at birth, compared to those that are *precocial,* or relatively mature at birth. Precocial mammals include hooved species such as the white-tailed deer. Despite being altricial, the raccoon cub's limbs are well developed and its whitish pink claws are sharp. However,

Mother raccoon with two day old cub. Photo by Julia Sims.

its helplessness does make it vulnerable to various dangers, such as preda-tors. This may be why its sparse hair is both directed forward and often imbued with a rusty color on the nape of its neck and shoulders. These features may provide the mother with a target to quickly grasp her young in her jaws and expedite their transport. When she perceives a threat, the mother grabs her cubs one at a time by the scruff of the neck, or occa-sionally around their middle, and totes them to safety. A distressed mother will often move her litter to a new den.

The raccoon mother usually remains with her cubs for their first night and sharply decreases her activity for the next few nights. Ordinarily, rac-coon mothers strictly curtail their movements for a few weeks, though re-ductions in their home ranges are not always apparent. Before too long, she resumes her nightly foraging trips but will still spend each day in the den with her young. The mother-young unit is the most common type of raccoon social group.

While the infants are restricted to their den, they will only consume milk; their mother does not bring them any solid food. At first, the mother will lie either on her side or back to nurse. As the cubs grow, she will sit up and hold one or more up to her teats. The cubs may rub the teat, per-haps to stimulate milk flow. Raccoons usually nurse for 8 to 16 weeks, though some may stop sooner while others continue to nurse for several months longer. As long as they are nursing, the mothers neither ovulate nor come into estrus. *Weaning*, the cessation of nursing, begins when the cubs leave the den and begin to forage on their own. The onset of this behavior is variable.

The following growth data were recorded from two captive raccoons: on average, a cub weighed 196 g at 7 days, 454 g at 19 days, 567 g at 30 days, 681 g at 40 days, and 908 g at 50 days (454 g = 1 pound). From the time they are weaned until late autumn, a cub's weight may increase by almost a 1000 g (2.2 pounds) a month. By the fall, a young raccoon can weigh up to 7 kg (15.4 pounds).

It has been reported that raccoons do not reach full growth until their second year. Young raccoons in Alabama, for example, usually attain adult size during their second autumn. Those from northern populations, how-ever, may not reach adult size until after their second year or later, per-haps due to the severe winter conditions they face. Male and female growth rates may differ. In south Texas, the males grew more rapidly than the females did during their first year. Interestingly, though, the females there may attain adult size during their first winter, whereas the males did not reach their adult size until they were almost two years of age.

Infants that are less than three weeks old squirm about rather than walk because their legs cannot support their weight. Captive raccoons have been observed to start walking between their fourth and sixth weeks. By the end of their seventh week, they can run and climb with some skill. Their eyes then react to objects and motion, and they regularly play with each other. Throughout this period, the mother is actively involved in their learning, encouraging or scolding them as necessary. The cubs may leave the natal den at this time, but do not regularly accompany their mother on nightly foraging trips until their 11th week, an activity that may start between their 7th and 12th weeks. In Texas, for example, the cubs began to travel with their mother and use other dens at an average age of 10 weeks; in Minnesota, this occurred at 11 to 12 weeks. At first, the cubs will start to follow their mother but then soon retreat to the safety of their den. Over the next few weeks, they gradually increase their distance from the den. A mother may move her cubs to a ground den when they begin to move freely from the tree den. This could be done to prevent their falling from the tree or perhaps to prompt them to find their own food. In one study, litters were relocated to the ground when the young were 45, 47, and 63 days old.

In most areas, juveniles become more fully independent from their mother during either the fall or winter. In Minnesota, for example, young raccoons started to rest alone at four to five months of age. Family groups in Michigan were also suspected to separate in the fall. Presumably, the mothers went off on their own while the young generally lived either apart or in groups of two or three. In colder areas, a mother and her young may resume sleeping by one another when the nighttime temperatures consistently fall below freezing. After the first permanent snowfall, they may either den together within the same hollow tree or near one another throughout the winter. They then disassociate once again during the following spring. Such familial associations may persist longer in northern populations because denning together in the winter reinforces the bonds between a mother and her young.

In the southern states, juveniles often become fully independent from their mothers during their first autumn. However, in many of the studies that have reported this, the raccoon families may not have been carefully observed. The comprehensive south Texas study by Stanley Gehrt and Erik Fritzell revealed that family relationships can be maintained even where the winters are mild. Most of these cubs stayed with their mothers throughout their first fall and winter, sharing resting sites more than 75 percent of the time. They began to rest apart at a median age of seven months, with a range of five to nine months. Their bonds usually lingered

Juvenile raccoon, probably several weeks old. Photo by Julia Sims.

until either the young or their mother became sexually active in the following spring.

Sexual maturation might play a central role in the breakup of raccoon families. Therefore, differences in its onset might help to explain the varying lengths of family bonds. In Texas, none of the juveniles that associated with each other or their mothers through the winter were sexually active during what would have been their first mating season. In addition, no juveniles were observed with their mothers when the adults were involved in mating. After this period, two of the litters returned to share resting sites with their mothers, an arrangement that lasted until the mothers produced new litters. Hence, mother-offspring relationships in raccoons may occur on essentially a year-round basis in some places.

In this Texas study, the bonds between siblings lasted for about as long as those between the mothers and their young. The siblings remained associated with one another throughout their first autumn and winter. Then, when they were about 10 months of age, around the time of the mating season, their level of association decreased by 70 percent. Some of the siblings, usually sisters, continued to share resting sites at this time, but otherwise they did not associate. The connections between yearling siblings remained quiet into the summer, and by autumn they had ceased. After mothers with their young, the next most common type of raccoon group seems to be yearling pairs. These may remain together for a short time following the dissolution of the mother-young bond.

After the bond between a mother and her cubs is severed, the young raccoons often leave the area in which they were raised. This departure from one's birth area is known as *natal dispersal*. Although aspects of this movement, such as the age at which the young leave or the extent to which they travel, are highly variable, some patterns in this activity appear consistent. Primarily, dispersal events can differ sharply between the sexes. Male raccoons leave the natal area earlier, are more likely to disperse, and typically travel for longer distances than the females.

Males may leave in their first autumn, though they might also disperse during the following spring or summer. Dispersal may occur in pulses; it does not necessarily take place at any one time, even among those in the same area. In North Dakota, the yearling males that dispersed in the spring and summer traveled between 0.3 to 10 km (about 0.2 to 6.3 miles) per night. Ultimately, most probably moved more than 20 km (12.5 miles) away, a distance comparable to that of the juvenile males that dispersed in Michigan. The males in Texas also largely dispersed in the spring as yearlings: 80 percent left when they were at least a year old, and virtually all of the young males eventually moved away. Some males travel significant distances both while dispersing and afterwards. In Minnesota, one tagged as a juvenile was recaptured three years later 264 km (165 miles) to the north. Another juvenile, marked in the late spring in Manitoba, Canada, was killed 253 km (158 miles) away less than six months later. Why or how often such long-distance relocations occur is not known, though they seem to be rare.

If female dispersal does occur, it is apparently more common in the spring among the yearlings. In a Michigan study, as stated, the family groups broke up in the fall. These females' average dispersal distance (13.2 km [8.3 miles]) was less than that of the males (18.9 km [11.8 miles]). In both North Dakota and Texas, however, the young females did not disperse. In the latter study, at least 11 of the 14 females that were more than a year old were closely related to a resident female. The associations between the female relatives in Texas seemed to persist, as indicated by the considerable degree of home range overlap among those at least 22 months of age. Three females even shared home ranges with their adult daughters and several shared communal den structures. In one remarkable case, the females from three generations of one *matriline* (a line of female descendancy) simultaneously raised their young in the same brushpile. Though the adult females there were ordinarily unsociable, related ones would sometimes associate in the evenings. Such tolerance of females toward their adult relatives has been reported in other solitary mammals, including the black bear.

Raccoon dispersal patterns correspond with the predominant mammalian trend that most males tend to leave their original homes, whereas females tend to be *philopatric*. The term *philopatry* is defined as the propensity of young individuals to remain in their natal area. In some species, dispersal may be provoked by competition between parents and their young for resources such as food or space. However, it is doubtful that such competition would be a factor in raccoon dispersal. In the Texas study, the researchers did not notice any negative interactions between the young males with either their mothers or fathers before they dispersed. Nevertheless, the adult males formed social bonds with each other but not with the young males. This may have effectively forced the young males to leave so that they would be unable to compete with the adults for mates. Their dispersal could also be explained by the *inbreeding avoidance* theory: Young males might leave an area to avoid mating with a relative. As discussed in the next chapter, on the south Texas study site, a male that remained in his birth area would end up sharing it with many closely related reproductive-age females.

Female raccoons may be permitted to remain in their natal areas because mothers and their daughters probably do not compete much for either resources or mates. Given their opportunistic foraging habits, not many of the foods that raccoons exploit are likely to become significantly depleted if a few more were to stay in an area. Furthermore, within their primary mating systems, *promiscuity* and *polygyny* (which are described in the next chapter), significant competition among the females for mates may be unlikely to occur. Therefore, a mother's reproductive success should not be jeopardized if she allows her daughters to live nearby. In fact, her overall *fitness* (the long-term success of her genes over generations) would be enhanced if her daughters' survival and productivity were improved by them remaining close. Surely, the young females should benefit from being able to stay in a familiar area, given that so many of the raccoon's foods are transitory.

Obviously, these explanations for why the females can remain in their natal area again beg the question of why it is that the young males tend to leave. Even though significant competition for mates might also seem to be unlikely among males, they actually may be more affected by mate competition in some areas, as evidenced by their exclusion from the adult male groups in Texas. How the raccoon's social behaviors affect its reproductive success is explored in the next chapter.

9

Social
Organization

Many factors can determine a species' social organization. Among
these, environmental influences such as the habitat type and the
distribution of resources in an area often play a critical role. But
because the raccoon occurs in so many kinds of habitats, explanations for
how the environment affects the ways in which males and females relate
to one another are neither simple nor consistent. Its home ranges, how
the location of an area's resources affects its spatial organization, its mating systems, the roles of sex-based differences in certain traits, and ultimately how these variables are interrelated determine the raccoon's *social
ecology*.

HOME RANGE SIZE

As mentioned in Chapter 3, a home range is the area an individual uti-
lizes on a regular basis during a given period, such as a particular season.
It is an important factor to consider in regards to social biology because
it can help us understand why the spatial patterns of males and females
vary in relation to one another. As is true for other species, many factors
affect the raccoon's home range, including the individual's sex and age, its
metabolism, the density of the population, habitat features such as the
distribution of food and water sources, and surely the interplay of such

influences. Despite the great variation in their home ranges, several consistent patterns have emerged from studies of raccoon populations, some
of which are predictable based on findings from other animals. First of all,
males typically have larger home ranges than females, which is to be expected because they have a larger average body size. Not surprisingly,
adults have larger home ranges than juveniles, and those of females with
young are usually limited. Last, in relatively northern climates, a raccoon's
home range tends to decrease in the winter.

In early field studies, usually only the diameter of the raccoon's home
range was estimated. Though diameters were commonly reported to be
from 1 to 3 km (0.6 to 1.9 miles), both considerably smaller and larger
ones are now known to occur. Like other animals, raccoons tend to have
smaller home ranges when their densities are high. Thus, in one crowded
suburban population, their home ranges were just 0.3 to 0.7 km (0.2 to
0.4 miles) in diameter. In uncrowded areas, they can be considerably
larger. On North Dakota's prairies, for example, raccoon home range
diameters of up to 10 km (6.2 mi) have been observed.

The raccoon's actual home range areas have only recently been determined, primarily through the use of radiotelemetry. In this technique, a
radio transmitter is affixed to an animal so that an investigator can follow
its movements with a receiver. Although it has been suggested that their
home ranges typically extend from 40 to 100 ha (99 to 247 acres), a wide
variety of values have been discovered. As suggested, population density
and habitat openness influence home range area. These vary from less
than 5 ha (12 acres) in Ohio's suburbs to the almost 5,000 ha (12,355
acres) home ranges of the adult males on North Dakota's prairies, a thou-
sandfold difference! In south Texas, Gehrt and Fritzell found that adult
male home ranges varied from 200 to 931 ha (494 to 2,301 acres),
whereas those of the females ranged from 14 to 535 ha (35 to 1,322
acres). These comparatively large home ranges occurred in a subtropical
environment, which obviously differs from the temperate areas of many
raccoon populations.

As stated, adult male raccoons commonly have larger home ranges than
adult females; this is the case throughout the year. This difference could
simply be due to the fact that males are larger and thus require more space
to search for the additional food they need to satisfy their energy requirements. During the mating season, which may occur in the winter
and spring in some areas, the males may further expand their home ranges
as they pursue females. Each may seek opportunities to mate with several
females, which could increase their travels even more.

In south Texas, however, the adult males did not expand their home ranges during the breeding season; their ranges remained consistent in size throughout the year. Interestingly, though, their home ranges were much larger on average than the area predicted from the overall correlation between body mass and home range size in mammals. This implies that a male's home range in that region is greater than the area needed to satisfy its energy requirements. This phenomenon could have several explanations that could apply to other areas where males have such disproportionately large home ranges. First of all, the males in this region may need larger home ranges throughout the year because of food and water source distributions, and so they do not expand them further during the mating season. Furthermore, traveling over large areas throughout the year might improve their chances of mating with the likely greater number of females within them. This could be an effective mating strategy if it does not cost them a prohibitive amount of energy.

TERRITORIALITY AND MATING BEHAVIORS

There are conflicting reports about how much the home ranges of adult males overlap. This is probably due to differences in the raccoon's social organization from one area to the next. Whereas some studies suggest that their home ranges broadly overlap, others indicate that there is very little overlap, with most individuals staying at least 2 km (1.2 miles) apart. The issue of home range overlap might be related to how (and if) the males are expressing their *territoriality*, which is technically defined as the active defense of an area. One would expect comparatively little overlap if they were defending sizable territories. Although no compelling evidence indicates that either males or females actively defend areas, both sexes exhibit signs of being territorial. For example, individuals of each sex tend to remain in distinct areas, even where their home ranges overlap. Males may maintain areas for their own use by mutually avoiding one another, and some dominant males protect temporary feeding territories. Encounters between females seem to be rare, as well, though some will congregate at feeding sites and engage in communal denning.

Yet another indication of territoriality in adult males is that each may maintain priority access to an area resources, such as its food or females, by being socially dominant. One study found that males trapped relatively close to each other displayed a greater frequency of dominant-subordinate relationships yet fewer intensely aggressive interactions than those caught farther apart. In another investigation, a male was observed driving away the other raccoons that approached the female he was accompanying. It may be that neighboring raccoons both recognize and avoid each other. These observations indicate that a fundamental social structure exists even when the individuals are acting in a largely solitary fashion.

Gehrt and Fritzell's study of raccoons in Texas is worth reviewing in detail because it is one of the most comprehensive investigations of their social organization. In this study area, the adult females were usually solitary. Their home ranges varied throughout the year, although not in concert with each other. Not surprisingly, they focused their activities in areas with abundant water or food. The sites with water in this region are often widely separated and ephemeral due to recurrent droughts. When standing water was abundant and uniformly distributed, the females were not particularly concentrated. But when the water sources became restricted to just a few widely separated patches, they became highly aggregated. Even when the females' home ranges overlapped extensively, they tended to forage and rest apart from one another. Though some seemed to feed and rest together, this was probably coincidental, and they moved about independently of one another. Thus despite appearing to be tolerant of other individuals, these adult females rarely associated with each other unless certain resources were restricted, and those that did typically were related. Such connections between female spatial patterns and the distribution of resources have been documented in several other species.

The adult males in this Texas study tended to occur in groups, typically of three to four individuals, throughout the year. The collective home ranges of adjacent groups overlapped little and accordingly, there was a lack of interaction between the males in them. Their separation might have been due to some form of territorial behavior. Similar male coalitions occur in other Carnivores. The organization of these groups could also be analogous to that of the individual males in a North Dakota population. These males' home ranges also did not overlap, and it was argued that they behaved territorially to maintain exclusive access to the females in their areas. As mentioned earlier, the densities of this North Dakota population were rather low (0.5 to 1.0 raccoons per km^2).

As the number of females in an area increases, a male's ability to maintain sole mating privileges with them typically declines. Thus, where densities are higher, as they were in Texas, males might be inclined to form coalitions to enhance their mating opportunities. Whereas in North Dakota the individual male home ranges usually encompassed those of two or three females, in Texas the shared home range of three or four males overlapped the smaller home ranges of at least 12 females during the mating season.

Evidently, the male groups in Texas formed in response to the repeated clustering of adult females at widely separated water spots. This situation is consistent with the general mammalian pattern of female distributions being affected by resources, and male distributions being affected by the distributions of the females. It also seems that the male groups were maintained by the interactions of their members. Those within a group frequently interacted and a dominance hierarchy seemed to develop, as reflected by their varied mating success. Several solitary males were also in this population. They occupied areas without females and displayed little or no overlap with the males in the nearby groups. During the mating season, they would temporarily move into neighboring areas, presumably because females in reproductive condition were present. Such circumstances have also been observed in solitary members of the weasel family.

The above studies and several others strongly suggest that raccoons tend to have either a polygynous or a promiscuous mating system, or some combination of the two. In a polygynous system, a male mates with at least two females. Various forms of polygyny exist, ranging from relatively loose arrangements in which males mate with a number of females seemingly at random, to more organized structures such as those in which a male actively defends a harem within a defined space. Certainly, the raccoon engages in a loose form of this mating system. In fact, its social structure seems so loosely organized in some populations that it has been described as promiscuous. In promiscuous mating, males and females may each couple with various partners throughout the breeding season. This often occurs in such a haphazard manner that it is difficult to even characterize this behavior as belonging to a particular system. Again, the raccoon's mating arrangements may vary between the two systems, even concurrently within the same area, depending on the degree to which a male, or perhaps a group of males, has exclusive mating rights to the females in its home range.

Thus, the variation in a male raccoon's mating behavior, and hence the mating system in an area, appears to be a function of the number and distribution of the adult females present. Given the wide variety of circumstances that they encounter, it should not be surprising that males will engage in a number of "strategies" to increase their reproductive success. In contrast to the situation in Texas, adult females are typically widely dispersed at low population densities, such as those occurring in North

Dakota. The males in this area may maintain separate home ranges to gain breeding access to the reproductive females in them, and may modify these areas so that they encompass the several smaller ranges of the females.

In North Dakota, at least two adult females exclusively occupied an adult male territory, indicating that the raccoons in this area have a polygynous mating system. At many southern latitudes, however, their food resources are more regularly abundant and thus their densities tend to be higher. When this occurs, the home ranges of the adults of either sex are likelier to overlap extensively. An additional influence on the raccoon's distribution patterns in Texas was the aforementioned changing nature of the water sources. This apparently caused the adult females to aggregate in different ways throughout the year, which the adult males reacted to during the breeding season. Based on the degree to which the male coalitions were able to retain exclusive breeding rights to the females in their territories, the mating system alternated between polygyny and promiscuity.

In addition to the females' distribution patterns, certain aspects of their reproductive biology may affect the males' breeding fortunes and thus the kind of mating arrangement. Within the raccoon's type of polygyny, a dominant male can only *sequester* or isolate one female at a time while attempting to mate. Apparently, the timing of the period of estrus or sexual receptivity among the population's females can affect a male's success at such efforts. Gehrt and Fritzell suspected that variation in the consortship success of their population's males would lessen with an increase in the *synchrony,* or simultaneous occurrence, of estrus among the females. This is because they assumed that the population's subordinate males would be able to gain greater access to its females if more of them were sexually receptive at the same time.

Indeed, the degree to which that area's females' estrus periods were synchronized had just such an effect on the raccoons' mating patterns. At those times when the synchrony of estrus was greater, the variation in the number of consortship days among the males decreased. This suggested that access to estrus females had increased for the subordinate males. It is important to recognize, though, that while such consortships do reflect a male's access to a receptive female, the precise connection between consortship and actual mating success is not known. Clearly, each of the dominant male raccoons has a vested interest to "push" the mating system toward polygyny so that he may secure more of the matings for himself. Nevertheless, several factors can affect his success at doing this, such as the distribution of a population's females, the synchrony of their estrous cycles, and perhaps multiple matings by the females as well.

Although some males commonly mate with several females each spring, pair bonds between individuals may still occur in some areas. On the other hand, though a male and a female may even den together throughout the winter and bond with one another a month before mating, the female may still breed with several males. After the mating period, no associations between males and females are apparent, and the males provide no assistance in rearing the cubs. Therefore, a male's reproductive success is primarily determined by his ability to successfully mate, often with several partners. This also helps to explain the discrepancy in male and female home ranges.

SEXUAL SELECTION AND DIMORPHISM

Having established that there are consequences of the male raccoon being larger than the female, such as having a more sizable home range, a more fundamental question might be asked: Why should the male be bigger? The relatively larger size of male raccoons and other traits that are discussed in this section are likely due to the effects of a process known as *sexual selection*. Charles Darwin, arguably the most influential biologist of all time, first described this phenomenon in 1871. During this type of selection, certain physical and behavioral traits are "selected for" or "against" based on the advantages or disadvantages, respectively, that they confer upon an individual's potential reproductive success. Sexual selection is regarded as a kind of natural selection, the central mechanism of evolutionary change first described by Darwin and his contemporary Alfred Russel Wallace in a joint presentation to the Linnean Society of London in 1858. The principal tenet of natural selection is that the traits that are selected for should be advantageous to an organism's fitness, or its survival to maturity and its reproductive success.

Two basic types of sexual selection exist. The first, *intrasexual selection*, entails interactions among members of the same sex. For example, competition between males for mates may affect the expression of such traits as their larger size or the types of body weaponry they use for fighting. The other major form of sexual selection is *intersexual selection*, which is based on interactions between members of the opposite sex. A common example of this process is how a female's choice of mates acts so that particular features in males are favored. Thus, females may prefer to mate with big males or those that have large antlers, and such traits are favored by this type of selection. Of course, it is difficult to determine precisely why a certain feature has evolved. One could readily presume that structures such as antlers have been and still are influenced by both intrasexual and intersexual selection.

One of the predictions generated by sexual selection studies is that the traits that are used in male-male combat should exhibit a high degree of *sexual dimorphism*. This term refers to differences between males and females in the size or form of a particular characteristic; those demonstrating such contrasts are called sexually dimorphic traits. Larger size in one sex is such a trait. Therefore, according to the previous prediction, male raccoons should be relatively large if this would be beneficial in fighting to protect territories and perhaps in guarding reproductive females from their rivals. In addition, a female might be hostile toward a male trying to copulate with her. If this occurs, a larger male would have an advantage during mating by being able to control such an uncooperative mate. In a roundabout way, this might be viewed as a skewed version of the female choosing her partner; by not being able to prevent such a mating, the relatively sedentary females essentially "select" for large size in the males by default. Some female choice might also be involved in discriminating among the several males that are attracted to her while she is in estrus. Thus, both intersexual and intrasexual selection pressures might be invoked to explain the larger size of male raccoons. Similar interpretations for sexual dimorphism have been offered for size differences in weasel species.

Biologist Mark Ritke has reported on several other aspects of sexual dimorphism in the raccoon. Males also have proportionately larger lower canines than females, and not simply because they have larger skulls. Males might use these teeth as weapons in fights with other males in mate competitions. Furthermore, a male's canines may have evolved to be wider than those of a female because of the strain they experience while fighting; wider teeth should be less susceptible to breakage. Such reasoning has also been used to explain the relatively large canines of certain male primates as well as the thicker horns of various male hooved mammals. Therefore, intrasexual selection may be at least somewhat responsible for the larger and wider canines of male raccoons.

Males also tend to have broader *rostrums*, the snout area of the skull. The greater width in this region may help brace the roots of their upper canines, which also may be comparatively larger than those of the females. In addition, males have wider *zygomatic arches*, the curved bony extensions that emanate from the sides of the skull. These serve as places for the attachment of certain jaw muscles, the *masseters*. Wider arches may accommodate larger masseters, which in turn produce a stronger bite, another capability that is useful in male-male combat. The arches also shield the blood vessels and nerves underneath them, so wider ones may lessen the likelihood that a bite to the head will result in damage to these structures. Therefore, intrasexual selection may also influence the width of a male raccoon's zygomatic arch. Defense against predators probably does not play much of a role in these differences between males and females. As indicated, adult raccoons have few predators and the males do not protect either their mates or their offspring.

Finally, the value of another concept called *niche partitioning theory* to explain the differences in size and other features of male and female raccoons is questionable. According to this theory, the traits involved in feeding should exhibit sexual dimorphism if males and females significantly compete for food. However, no evidence indicates that male and female raccoons employ different feeding strategies, eat dissimilar foods, or compete in any way for food. Rather, the likely explanation for the raccoon's sexual dimorphism is that it results from male-male competition that selects for larger size and other structures in males, and perhaps the indirect effect of female choice for larger mates. These differences in male and female characteristics coincide with the raccoon's tendency to have promiscuous and polygynous mating systems.

10

Management

The management of a species largely involves learning a sufficient amount about a particular population so that recommendations can be made for it based upon established objectives. In most areas, raccoon populations are managed so that a certain number of individuals can be harvested by trapping and hunting without affecting the population's capacity to sustain itself. First, researchers must determine the basic biological features of a raccoon population, and then decide how such information can be used to manage it. One reason to manage a raccoon population may be to reduce excessive predation on different species, particularly various birds, as mentioned earlier. Another reason may be to control rabies outbreaks. As discussed, raccoons are significant carriers of rabies, so much so that in many areas this has become a source of alarm. Because of these problems and simply because they are perceived as pests, the raccoon is managed as a nuisance animal in many locations. Thus, raccoon management also involves their removal and *translocation*, or relocation. However, several raccoon species, notably the island forms, require special attention to enhance their numbers if they are to survive. (Issues pertaining to their conservation are discussed in Chapter 3.)

AGING AND SEXING TECHNIQUES
AND THEIR IMPLICATIONS

Understanding the changes that occur in the number of individuals in a population through time requires a knowledge of its *age structure*. This term refers to either the numbers or relative proportions of individuals in each age class, such as the juveniles, yearlings, and adults in the population. If a population is increasing, it is likely to have a comparatively high percentage of individuals in the younger age classes because these contribute to its future growth. The reproductive output of those from different age classes may likewise be an important determinant of population change. Knowing the number of individuals of each sex within the population and their levels of mortality is also crucial. Obviously, a multitude of factors must be considered to gain a full understanding of a population's dynamics. Therefore, to even begin to assess population trends and hence ultimately provide sound management recommendations, biologists need to know how to determine a raccoon's age and its sex.

The most basic age distinction one can make is that between juveniles and adults. If these groups are separated on the basis of their ability to reproduce rather than their actual ages, these data can provide information about the proportion of reproductive individuals in a population and therefore its potential productivity. In such analyses, subadult (yearling) females are considered to be sexually mature, yet subadult males are regarded as immature. This is because such females typically contribute to a population's productivity whereas the males do not. Note, however, that yearling females generally do not contribute to a population's productivity to the same degree as adult females. In addition, though yearling males may be capable of reproduction, they are usually not part of the breeding group.

Many criteria have been employed to separate raccoons into juvenile and adult age classes. These include using the size of the pelt, body weight, eye lens weight, and canine sharpness. The use of any of these measures, however, can result in a large percentage of the population being erroneously classified. For example, Monique Kramer and her colleagues determined that aging techniques involving eye lens weight and tooth wear accurately categorized almost all of the sexually mature males in their study. Yet using canine sharpness measurements resulted in more than a fifth of the sexually immature males being classified as mature. Another technique used to distinguish juveniles from adults is based on the fact that as a mammal ages, the cartilaginous areas between sections of certain bones close up as they are replaced by bony material. These

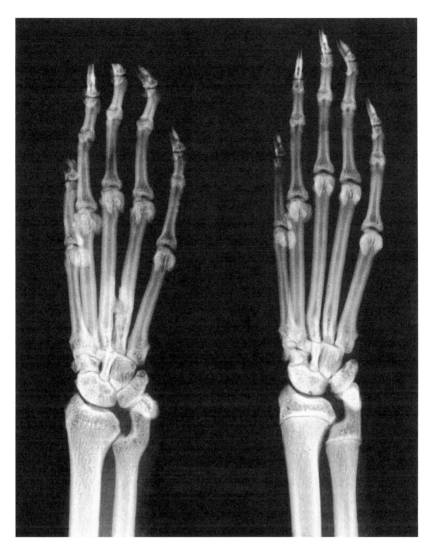

x ray of the wrist areas of a juvenile (right) and an adult raccoon showing differences in the closure of the epiphyseal suture (under the top knobs of the limb bones). Note how the space of this suture is open and thus more evident in the juvenile. x ray by Samuel I. Zeveloff; photo by Christopher Nutting.

spaces, referred to as suture areas or *epiphyses* (*epiphysis*, singular), can be examined visually in cleaned bones or with x rays. Closure of these epiphyses in the raccoon's wrist area has also been used to differentiate juveniles and adults.

To distinguish juvenile from adult males, as well as sexually immature from mature ones, researchers have also considered the *extrusibility* of the penis. This term refers to whether or not the penis can be advanced through an opening known as the *preputial orifice*. The penis of a juvenile raccoon is not extrusible. Yet in their analysis, Kramer and her colleagues found that even though the use of this measure resulted in classifying less than 2 percent of the sexually mature males as immature, more than 20 percent of the sexually immature ones were categorized as mature. The development of the baculum, or penis bone, can also be used to distinguish juvenile from adult males. In juveniles, the baculum is small and the tip at its far end is considerably more cartilaginous. Those of adults are heavier and longer, and have a gradual curve near their enlarged base.

Females can be examined externally for evidence of pregnancy or lactation, and thus their sexual maturity. Pregnancy may be determined simply by feeling if it has progressed far enough. A female's teats often become enlarged and darker during pregnancy and nursing, and therefore may also be used to assess her breeding condition. In the study by Kramer and her colleagues, however, the use of these seemingly reasonable measures resulted in the incorrect classification of many of the mature females. This could have occurred because some females had only recently bred and did not yet display signs of maturity, and because teat color does not change in every sexually mature female. The maturity of dead and anesthetized females can be also discerned by examining their reproductive tracts for uterine swellings, fetuses, placental scars, and the presence of *corpora lutea* (*corpus luteum*, singular) in the ovaries. The latter structures, which form where the eggs are shed, are indicative of pregnancy.

Within the past few decades, other procedures have been developed to group raccoons into several age classes other than just that of juveniles versus adults. For example, the pattern of tooth eruption may be used to age preweaned raccoons. Although such information is probably not useful for managing them, knowing the ages of preweaned individuals might help to elucidate patterns of juvenile mortality and growth. The sectioning of teeth to show their *cementum* layers (the material on teeth that secures them in their sockets) allows raccoons to be separated into eight age classes: young of the year; those at one, two, three, four, five, and six to seven years; and those more than seven years old. The last method of aging raccoons is by *cranial suture obliteration*, an approach that is analogous to examining the closure of their epiphyseal sutures. The various bones of the skull fuse with their adjacent ones in a certain sequence. By evaluating which cranial sutures have become indistinct or obliterated at successive times, a raccoon's age can be determined at 2-month periods up through their 50th month. Obviously, most of these internal aging crite-

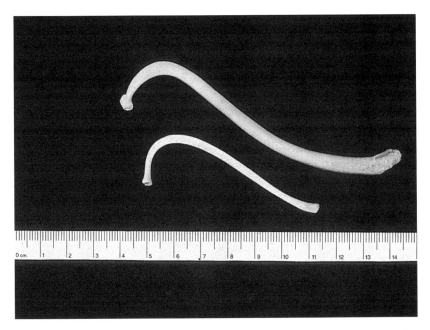

Bacula of adult (top) and juvenile raccoons. Though considerably more cartilage is normally on the tip (left) of a juvenile's baculum, this material has been removed. Photo by Christopher Nutting.

ria can be applied only after an animal has died. In some areas, obtaining a sufficient number of carcasses to provide reliable estimates of the numbers in each age group may be difficult.

Despite the problems with the reliability of these procedures, biologists still commonly use them to gain an understanding of a population's age structure. For example, Glen Sanderson's long-term study in Illinois revealed that young of the year accounted for nearly 70 percent of the total harvest. In general, though, the proportion of juveniles seems to vary greatly in northern areas, where they have been found to range from as little as 20 percent to as much as 70 percent of the raccoon population. The proportion of juveniles generally appears to be lower in the southern United States. In an Alabama study, for example, juveniles composed only 32 percent of the population.

This seeming contrast in age structure could be related to differences in productivity as well as to the greater mortality that raccoons suffer from severe winters and often intense harvesting in the northern parts of their range. This explanation may seem illogical because one might expect harsh winters to exert their most intense effects on the juveniles, thus resulting in relatively low proportions of them in northern populations.

However, the overall higher winter mortality in northern climes could also result in diminished breeding populations, and the fewer remaining individuals may produce more offspring in response to their low density. Such an inverse relationship between breeding population density and the number of young has been documented for several species.

As suggested, wildlife managers are also interested in determining the sex of a population's individuals. Recording data at different intervals allows them to assess whether the population's sex ratio has changed and why. A raccoon's sex can be easily determined, from both living or dead individuals. The testes (which are always descended), the baculum, and the preputial orifice distinguish the male; alternatively, the female has a vulva and other reproductive traits. Even in an embryo's early stages, sex can be determined in males by the baculum's outline and the greater distance between the anus and the urinary opening. At the fur houses where hunters and trappers sell their quarry, pelts are stretched on frames or boards to be cleaned and dried. When these *cased* pelts are examined, a raccoon's sex can be determined by the location of the preputial orifice or the vulva. If the pelts are ripped around these openings, sex can still be established by the presence of the baculum or other identifying features in the carcass. One can usually locate and identify the ovaries and the uterus, even in small females.

In the aforementioned Illinois study, the relative numbers of males and females from before birth up to two months of age indicated that their sex ratio did not differ significantly from 50:50. When raccoons are more than about two months old, males may have higher mortality rates. As raccoons become mobile, the juvenile males may be more active than the females. Because these young males range over larger areas, they could be more vulnerable to certain mortality factors.

ESTIMATING NUMBERS AND POPULATION TRENDS

The techniques used to estimate numbers of raccoons and their population trends are largely imprecise. However, they can generate valuable indexes that compare population sizes at different times and therefore help determine population trends. Suitable methods for accurately determining raccoon numbers on a large scale are not feasible, as they often are for more readily visible species. Yet obtaining a reasonably reliable estimate of their population sizes is still possible. Moreover, the information that is obtained about their trends may be sufficient for their management over extensive areas.

Scent-Marking Indexes

This technique involves using scents to attract individuals to *scent stations*. These stations are typically composed of sifted soil to which a liquid scent is added. Scents are created from synthetic fatty acids, as well as red fox and bobcat urine. These stations are established at intervals, and the raccoon's seasonal use of different habitats is taken into account for their placement. Based on the number of visits to these, researchers can obtain a measure of an area's raccoon abundance.

Spotlight Surveys

In this type of census, a driver and an observer travel along routes in the evening using powerful spotlights to locate raccoons. Vehicles are driven slowly while the driver and observer count the raccoons on each side. These surveys tend to be more effective during periods of high humidity, low wind, above-freezing temperatures, and before the new spring leaves make spotting individuals difficult. Thus, they are less useful in southern states that have foliage throughout the year.

Fur Buyers' and Sealing Reports

One of the oldest methods of keeping track of raccoon populations is to use fur buyers' reports of the numbers harvested. Yet, as mentioned previously, such data may not truly reflect long-term population dynamics as socioeconomic conditions can affect the marketing of fur and the numbers harvested. Furthermore, those who sell raccoons to fur buyers might not remember their quarry's location, keep records, or submit reports. Also, the buyers may not correctly list prices or numbers, and could include raccoons in their counts that were purchased from others who have already included them in their own counts. *Sealing reports*, used in some Canadian provinces, similarly provide indexes based on the number of harvested individuals. The term *sealing* refers to distinguishing a pelt with a mark, which is required in some areas before it is sold.

Field Trial Surveys

Hunter success during field trials is also used as a method to estimate raccoon numbers. During these nonharvest events, points are scored for observing and treeing raccoons. In South Carolina, the organizations that host these events are required to share their results with the state's Department of Natural Resources. Data on the number of individuals seen per hour are then converted into an index of raccoon abundance. A considerable amount of information can be generated by this method at little cost to an agency. Furthermore, the hunters may be more apt to accept an agency's decisions after being involved in management-related activities.

Mark-Recapture

This technique, which includes several variations, has been used to estimate the population size of many species. Rather than just furnishing an index of the number of individuals, it can generate a realistic approximation of the population's size. It relies on capturing individuals and marking each with a distinguishing item such as a colored ear tag. Each can thereby be identified upon recapture or subsequent observation. To estimate population size, a formula is employed that considers the number recaptured relative to the number initially caught. Investigators have also used marked individuals from hunters, trappers, and fur buyers. Although the counts of marked animals can be insufficient to generate valid population estimates, these analyses remain common. There is a variation of this method that addresses the problem of small numbers of marked animals. It involves feeding or injecting individuals with *radioisotopes* (radioactive isotopes, or versions, of an element), which allows their locations to be determined by examination of feces. Radioisotope tagging can reportedly provide highly reliable estimates of raccoon population size.

Direct Counts

Attempts to actually count all of the raccoons in an area are rare due to the time and money they require, and because the previously mentioned techniques are often adequate for assessing their numbers and trends for management. It may be feasible to fully count raccoons in certain situations, such as when they are in their winter dens or other restricted settings. These counts might not measure the population's density, however, because one would not necessarily know the size of the area that the raccoons inhabit when they are not denned. The methodology used and the local conditions can markedly affect the accuracy of these counts.

EVALUATING TRENDS AND REGULATING THE HARVEST

To soundly manage raccoons, it might not be efficient to only monitor changes in their numbers. It could also be important, depending upon the management objectives, to understand a population's reproductive potential and appraise the significance of its mortality factors. Productivity measures, as suggested earlier, may be derived from female reproductive data as well as from the sex and age composition of the harvest. Unfortunately, even long-term data may not provide adequate information for numerical trends to be discerned. Sanderson amassed such data on Illinois populations for nearly 30 years, evaluating key parameters such as litter size and the percentage of the population composed of young of the year. Rather surprisingly, this enormous data set did not reveal any insights into major changes within the population, perhaps because none had occurred.

Some researchers have argued that changes in yearling pregnancy rates can have a substantial effect on a raccoon population because of their potentially great variation. Yet even though these rates may fluctuate considerably, a significant correlation between their annual changes and subsequent shifts in population size has not been clearly demonstrated. It has also been hypothesized that pregnancy rates in some raccoon populations are *density dependent*. This means that when the density of individuals in an area is comparatively high, the percentage of pregnancies becomes lower. A large population would thus tend to exhibit depressed growth because of its decreased pregnancy rate and perhaps other productivity measures. Therefore, if a population was sufficiently large and the fall or winter harvest was relatively light, the pregnancy rate in the following spring would be expected to be low. This pattern has also been observed in other mammals.

Such responsiveness of raccoon populations to changing densities was alluded to in discussing how the proportion of juveniles in an area may shift. It was described how high winter mortality results in smaller breeding populations, and that the fewer remaining individuals may produce a relatively large number of young in response to their lower density. William Clark and his colleagues conducted a study in southwestern Iowa, however, that only partially supports the notion that raccoon populations are density dependent. In the 1980s, the yearling pregnancy rate in this area was low following a comparatively light harvest. Nevertheless, after a substantial subsequent harvest, the population did not rebound to its previous levels in the following year. Such an increase would be expected if the density-dependent effects were powerful enough for the population to respond rapidly to numerical changes.

In any event, a far better understanding of the effects of mortality on raccoon populations is sorely needed. As is made evident in Chapter 7, "Mortality and Disease," very few studies describe the relative contributions of mortality agents to their demographics. Undoubtedly, disease has the potential to become an important raccoon mortality factor. In parts of Illinois, for example, canine distemper may be capable of decimating their populations for years. Similarly, it is believed that rabies may have the potential to diminish their numbers in the northeastern United States by as much as 90 percent. Again, these appear to be the only two diseases that can seriously impact raccoon populations.

As is also discussed in Chapter 7, the major causes of raccoon deaths are usually hunting, trapping, and motor vehicle accidents. Trapping, though regarded by many as a cruel practice, is likely to remain legal in North America for the foreseeable future. Hunting and trapping can not only substantially impact raccoon populations, but the impact of other mortality factors may vary with the level of exploitation. For example, in a part of Iowa subjected to intense harvests, nearly 90 percent of the raccoon mortalities have been attributed to hunting and trapping. Yet in a lightly hunted Minnesota population, greater than half of their deaths were ascribed to other causes such as starvation and parasitism (although the climate in each of these areas could also influence these differences). Thus, the setting of harvest limits and seasons is typically a key component of raccoon management.

A few additional cases further illustrate the degree to which harvesting can affect raccoon populations. In Alabama's Wheeler National Wildlife Refuge, the number of raccoons taken exhibited a fourfold increase after a six-year cessation of harvesting. Excessive harvesting, both legal and illegal, has been regarded as the primary cause of low raccoon numbers in east Tennessee, and unlawful activity may be a significant problem in other states as well. The dwindling of their numbers in Ohio during the first half of the twentieth century was ascribed to both hunting and the summer *running* of hunting dogs (allowing them to pursue raccoons). The latter activity may especially impact raccoons born in the previous spring; families can be broken up, decreasing the young's chances for survival. Frederick Stuewer was also concerned about the effects of dogs in Michigan, arguing that they should be prohibited from "running on" raccoons before mid-September in this location. But as a Mississippi study recently suggested, summer hunting, even with dogs, may not necessarily be detrimental to raccoon populations if it is effectively managed.

Another recent study, also recounted in the chapter on mortality, illustrates the degree to which hunting and trapping can affect raccoon survival, and thus may address the capacity of exploitation for affecting their population dynamics. This is the research conducted by Gehrt and Fritzell on Texas's Welder Wildlife Refuge, where raccoons have been protected from harvesting since the early 1950s. As indicated earlier, they compared the survival of individuals from their population with those from the intensively harvested population in Iowa that Clark and his colleagues studied. They found that the annual survival of both males and females were considerably higher in the unexploited Texas raccoons. Yet apart from the harvest season in Iowa, the survival patterns within the two populations were similar. The survival rate differences are clearly attributable to the additional mortality from hunting and trapping.

Numerous factors can affect the success of hunting raccoons and thus the potential impact of this activity on a population. The fortunes of raccoon hunters appear to decline during cold or dry weather. In an Iowa study, though, more hunting and trapping occurred in years that had a high number of snow-free days. Other influences on harvesting levels in this area were pelt prices, and to a lesser degree, the amounts paid for other types of fur. Significant changes in pelt value often appear to have a strong effect on the numbers of raccoons taken, though the exact nature of this relationship is not always clear.

Some of hunting's effects on a population can be rather complicated. In the southern United States, juveniles may be especially susceptible to fall hunting as they could be dispersing by then and would be more conspicuous to hunting dogs. In east Tennessee in 1976, about 75 percent of the harvested raccoons were juveniles; it was the highest percentage reported for a southern population. This resulted in the prudent recommendation that experimental hunts should be implemented to determine which harvesting strategies might generate sustainable populations. Accurate monitoring techniques should help to ensure that harvest regulations are closely related to a population's abundance and structure.

Though raccoons can apparently tolerate heavy exploitation levels when their populations are large, sustained intense harvests can have a depressing effect on their numbers. But even when hunting and trapping seem to be their most significant mortality factors, the regulations that govern these activities might not always have a sizable impact on their populations. For example, in the early 1940s, Missouri established a bag limit of 10 raccoons and shortened the season from 46 to 31 days. The harvest sharply decreased from about 30,000 individuals to about 10,000 and stayed low for two years. Three years later, when the season was restored to its original length, the take climbed back to 30,000 and continued to rise. Yet Iowa during the same period had no bag limits, and the hunting and trapping seasons in this state remained the same for about 20 years. Despite these differences, the raccoon's population patterns in both states were strikingly similar. Of course, these populations or their regions might have differed in some fundamental way. Nonetheless, substantially dissimilar exploitation regimes evidently did not result in discernible differences in the raccoon's population dynamics in these areas.

Another illustration of the difficulty in evaluating the impacts of harvesting practices involves events in Illinois during the 1970s. At that time, the number of raccoons taken annually in this state increased more than fourfold, from about 68,000 to more than 310,000. Apparently, the population sustained this onslaught because the harvest persisted at the higher level for several years. Though the harvest eventually decreased by about a third in the 1980s, the key factor behind this decline was thought to be extremely cold weather during the hunting and trapping seasons. Again, seemingly significant variation in exploitation practices may not have had much of an effect on their numbers. Perhaps the populations were so high initially that they could sustain heavy losses, and annual productivity and juvenile survivorship may have been high. Thus, the impact of a substantial harvest could have been dampened and the effects of the dissimilar harvest rates might have been masked. The harvest's impact is also likely obscured when the individuals under consideration are pooled from a wide area, such as an entire state.

Clearly, it is difficult to determine the effects of the length, intensity, or schedule of a harvest on raccoon numbers. The scheduling of the harvest seasons may vary widely, even within the same area. For example, during about a 15-year period in Illinois, the hunting and trapping seasons ranged from around 30 to more than 90 days. Their starting dates fluctuated throughout November and the closing dates ranged from late December through February. Ironically, the highest harvest occurred during a particularly short season. Then, despite the season being slightly longer the following year, the harvest declined by more than 17 percent. Obviously, longer seasons do not necessarily result in more raccoons being killed. As indicated, factors such as the weather during the hunting and trapping seasons, the intensity of the previous harvests relative to the population's size, and pelt prices all appear to affect the numbers taken. Ordinarily, it is wise to restrict harvest seasons to the late fall and winter, especially in heavily hunted areas. Raccoon hunting seasons normally begin only after the juveniles are mature enough to care for themselves and they close before the breeding season starts.

As discussed earlier, several biologists have assumed that male raccoons are more vulnerable to hunting and trapping than females, and there is some evidence that supports this notion. For example, the adult males appear to have had comparatively lower survival during the intense harvest in Iowa. Again, a male's increased movements, especially during the mating season, could affect his survival because he will experience higher energetic costs, added exposure to late winter harvests, and a greater susceptibility to disease. Interestingly though, male and female survival appear to be similar when the harvest level is lower than about 34 percent of the total population. Findings from Texas and Mississippi studies were consistent with the view that male and female survival differences are reduced when harvesting is either light or not allowed. The behaviors that increase a male's vulnerability apparently do not significantly impair its survival at low levels of exploitation.

Additive is a term referring to a type of mortality factor that generates deaths in a population in addition to those expected from other causes. Alternately, a *compensatory* mortality factor is one that takes the place of a death that would have occurred in the population. Thus, if harvesting were compensatory, the death of an animal from hunting or trapping should substitute for one that would have occurred anyway, perhaps from disease or predation. William Clark and Erik Fritzell proposed that harvesting has more of an additive than a compensatory effect in Carnivores.

The Iowa study, however, suggested that the harvesting in this area was largely compensatory, and that most of this compensation took place through changes in mortality in subsequent seasons. As indicated, raccoon density is expected to decrease after an intense harvest, so the survival and reproduction of the remaining raccoons could be enhanced because more resources would likely be available to them. Yet, the non-harvest mortality of the adults in Iowa was fairly low, and the survival rate differences in the Iowa and Texas populations only occurred during the harvest seasons. Thus, the Iowa harvest may have actually had an additive effect. Clearly, evaluating this matter and drawing conclusions from comparisons in areas as dissimilar as Iowa and Texas is difficult. Unfortunately, there are no studies that lend themselves to a contrast of raccoon demographics in otherwise similar harvested and unexploited populations.

The exploitation of animals appears to act in either an additive or a compensatory fashion depending on the proportion of the population that is killed. When the level of harvesting is relatively high, it may tend to have an additive effect. Therefore, such an effect could occur in many northern and midwestern states where the raccoon fur market is often strong due to the higher quality of the pelts in these areas. Yet in various southeastern states, where the numbers taken tend to be lower, harvesting is likelier to have a compensatory effect. This is not necessarily the case throughout the South. For example, a substantial amount of harvesting occurs in Texas and Louisiana along the Gulf of Mexico. Certainly, the extent of harvesting and its impact in any given area varies. Exploitation may also have an additive effect during the harvest but a compensatory one on a long-term basis, as was suggested for the summer raccoon hunt in Mississippi.

Wildlife managers often try to base the percentage of a population that may be harvested on whether the number killed has an additive or a compensatory effect. Thus, in a region of Iowa, Clark suggested that compensation could occur when the raccoon harvest levels were 20 to 40 percent of the estimated autumn population size. Harvests were said to behave additively when they exceeded 40 percent of that level. Therefore, to ensure that exploitation was not detrimental to the population's sustainability, the maximum potential harvest rate (the *harvestable surplus*) was said to be 41 percent of the average preharvest population. It is difficult, however, to calculate sustainable harvest levels for the raccoon. Considerably more research is needed to determine the degree of exploitation that might permit such compensation to occur in their diverse populations.

To thoroughly quantify the effects of exploitation on raccoon populations, other issues such as juvenile survival must also be considered. A recent unpublished analysis in South Carolina indicated that the density of trappers in an area might affect the age structure of a population. In areas with a greater number of trappers, the raccoon populations tended to have more juveniles and were characterized by higher rates of population *turnover* or extensive change. Alternately, in low-harvest areas, the populations were similar in size but tended to be composed of older raccoons and had lower turnover rates. The lesson is that different harvesting levels may not have a marked effect on the total number of individuals in a population, but they might generate differences in its age structure.

HABITAT MANAGEMENT AND POPULATION ENHANCEMENT: A BRIEF PRIMER

The most critical goal of managing almost any species is to ensure that there is sufficient suitable habitat for its populations. Thus, it seems clear that the best way to maintain healthy raccoon populations is to protect and improve their habitats. Ample evidence indicates that they react favorably to an increase in den sites and food supplies. For the raccoon, however, determining what constitutes good habitat is complicated because they thrive in so many environments. (Their habitat ecology and descriptions of where they are likeliest to be abundant are discussed in the beginning of Chapter 6, "Living Arrangements.")

Historically, suggestions about their habitat management have focused on the importance of an adequate number of den sites. Based on his pioneering studies in Michigan, Stuewer advised more than 50 years ago that potential den trees should be protected when trees are removed, a recommendation that has since been echoed elsewhere. For that area, he proposed that there should be at least one or two dens every 6 to 8 ha (15 to 20 acres). Raccoon tree dens take a long time to form, perhaps even more than 100 years in some cases. The process involves not only the period of tree growth but also that of cavity development. Once decay induces its formation, an adequate cavity could still take 20 years to form, depending upon the climate, site, tree type, and position of the opening. Therefore, trees that lack sufficient cavities but that are damaged or diseased in such a way so that den sites could form should be protected, as should the other structures raccoons use for dens, such as brush piles and ground dens. Where natural dens are scarce, placing den boxes in wooded areas, especially near water, may help to promote their success. Both natural and artificial dens are likely to shield them from predation and aid in recruiting individuals into a population.

Raccoons are often successful in areas with large numbers of fruit-bearing plants, such as persimmon trees and wild grapes, and nut trees, particularly oaks and beeches. Stuewer was also an early advocate of decreasing the cutting of nut-bearing trees to enhance their populations. Protecting woodlands in agricultural areas from cutting, intense fires, and grazing is also likely to benefit raccoons. Given their affinity for water, the protection of streams, marshes, swamps, and areas supporting beaver colonies, as well as the creation of additional ponds or marshes near wooded areas should be valuable to them. Obviously, the establishment of refuges with their preferred habitat features should help to maintain breeding populations of raccoons, and these could also serve as sources for their dispersal into nearby areas. The raccoon's continued success can be ensured by protection of their favored habitats, though this must be coupled with appropriate harvest regulations.

Though it is costly, various states have *stocked* raccoons. Raccoon clubs buy individuals from other locations and release them in their hunting areas. Such efforts, however, may not improve their numbers if the habitat is unsuitable or if they are not managed wisely. Raccoon stocking actually may not be necessary anywhere in the United States, given their capacity for increase and ability to thrive in such diverse habitats. If the habitat is not suitable or if they are not native to an area, it becomes difficult to justify their importation. Nuisance animals may also be translocated, as is discussed shortly.

HARVEST RECORDS AND ECONOMIC VALUE

Though data exist on the numbers of furbearers acquired by fur trading companies since the 1600s, the raccoon's status over the past few hundred years is both sketchy and unclear. To gain a sense of the numbers trapped during the nineteenth century, researchers commonly relied upon this industry's records, particularly those of the famed Hudson's Bay Company. This fur trading company appears to be the only one for which fairly complete data sets are available. The legendary naturalist Ernest Thompson Seton used their records to analyze the North American raccoon harvest. During much of the nineteenth and early twentieth centuries, he reported that an average of 3,500 was taken annually. Fewer than 1,000 per year were harvested between 1821 and 1835, and there were only 10 years between 1865 and 1905 when more than 4,000 were taken. Seton described three peaks in the harvest: the largest occurred in 1867, a lesser one in 1875, and a third in 1899. He suggested, however, that the 1867 record of about 24,000 raccoons was probably the result of an unplanned accumulation of pelts rather than of a high abundance at that time. Though he estimated that about 5 million raccoons were in North America when the Europeans first arrived, this approximation is unsubstantiated.

Seton did not provide information on the average pelt values of raccoons. Reliable data on this has existed from some states only since the early 1920s. The first full accounting of their U.S. harvest did not take place until the 1934–1935 season, when about 388,000 were taken. Between the 1920–1921 and 1947–1948 seasons, their average pelt value varied from around five to nineteen dollars. As indicated previously, their populations are believed to have been comparatively low before about 1943. But by 1946–1947, the harvest surged to more than 1 million and gradually rose to close to 2 million by 1962–1963, perhaps indicating that their populations were also rising dramatically. The harvest declined through the 1960s though, dropping to less than 900,000 in 1967–1968. In the late 1960s, it once again began to increase. Throughout the 1970s, a close relationship seems to have existed between pelt prices and numbers harvested, though again, a connection between these factors is not always evident. By 1976–1977, the average pelt price ascended to 26 dollars. The harvest reached a peak of about 5.2 million raccoons three years later and stayed close to that level for the next few years.

By the 1983–1984 season, however, the average pelt price plummeted to about $5.50 and the U.S. harvest had dropped to about 3.2 million raccoons. It hovered between 3.2 to 4.7 million throughout most of the 1980s, dropping to 2.2 million in the 1988–1989 season. During the 1991–1992 season, the average pelt price remained low at about $5.80, and only about 1.4 million raccoons were taken in the United States. Throughout the first half of the 1990s, between 900,000 to 1.9 million were harvested annually in this country. Despite the large numbers that are killed by hunters and trappers, as well as by motor vehicles and disease, the raccoon has maintained high population levels. In recent years, this seems to be at least partially attributable to the decline in their harvest. As mentioned, it was estimated that there were at least 15 to 20 times the number of raccoons in North America by the late 1980s as there were in the 1930s. Moreover, as described in Chapter 5, the raccoon has substantially increased its range within the past 60 years.

The numbers of raccoons harvested in Canada have been comparatively small. As discussed, raccoons have a considerably larger distribution in the United States. Between the 1919–1920 and 1968–1969 seasons, there were never more than 50,000 taken annually in Canada. Raccoon harvests in this country increased to what was a peak at that time of almost 200,000 in 1981–1982. Usually, the greatest numbers of raccoons taken in the Canadian provinces are from Ontario. Records of the numbers harvested in Mexico and other countries are not readily available, and such data are not recorded by the International Association of Fish and Wildlife Agencies, the quasi-governmental organization that compiles harvest statistics of furbearers in the United States and Canada.

As stated, the raccoon is unquestionably the most economically important furbearer in North America. In the 1920s and 1930s, raccoon fur farming was common in the United States, but it was not profitable and eventually ceased to occur. On the other hand, hunting and trapping them has been enormously profitable. Income generated by sales of raccoon pelts reached a peak of $100 million in 1982. The majority of the pelts, at least in the 1970s and 1980s, were exported to Europe, especially to what was then West Germany; they were commonly sheared and dyed to be sold as imitation otter, mink, or seal.

Finally, in addition to their economic value as furbearers, raccoons may be useful, albeit in an unfortunate way, as monitors of the environment. Because of their relatively carnivorous habits, they can be biomagnifiers of certain toxic materials. This is because when such substances are in the environment, they tend to become concentrated in higher level consumers such as Carnivores. Indeed, heavy metals have been found to become concentrated in their skeletal muscle, liver, and hair. Raccoons in several states have been found to have elevated levels of mercury and lead. On the Department of Energy's Savannah River Site in South Carolina, their muscle and liver tissues have been analyzed for the radioactive substance radiocesium. Reliable data about certain types of pollution may be generated by as few as five raccoons per square mile.

RACCOON CONTROL

The raccoon is one of the most chronic nuisance animals in North America. It creates innumerable headaches for people living in vastly different environments. The damage that they cause and the efforts to move them generate substantial costs for individuals and government agencies alike. They are often the most common pest species in an area. In 1994, for example, raccoons accounted for almost 30 percent of the more than 45,000 mammals in Illinois handled by people with nuisance wildlife permits. Their predation on birds and their eggs, especially waterfowl, also causes considerable problems. Although their impact on bird populations varies, as described in Chapter 6, raccoons can have a severe effect on them in certain circumstances.

As reviewed previously in this chapter, an intensive harvest may generate additive mortality pressure on raccoons. Therefore, high harvest levels can be used as a management tool to reduce their populations. Apparently, trapping is usually the most effective method of raccoon control, as hunting may only infrequently have a significant impact on their numbers. This comparison, however, may be most applicable to declines resulting from typical harvesting seasons. In one attempt to deliberately lower a population on Alabama's Wheeler National Wildlife Refuge, hunters were able to remove about four times the number of raccoons that trappers could (although it is not clear whether a similar effort was devoted to each endeavor). Even when hunters have the potential to eliminate more raccoons, trapping may be more efficient as it should involve less time.

Many techniques have been developed to reduce the impact of nuisance raccoons. Usually, electric fences will keep them out of areas such as gardens or chicken yards. Nest boxes erected to benefit waterfowl, such as wood ducks, must be appropriately designed and spaced apart to minimize their predation. A recent test has revealed that raccoons reduce their consumption of bird eggs after being exposed to those treated with estrogen, apparently due to its taste. Poisons such as strychnine and Fumarin, an anticoagulant, may be used to reduce especially high populations. However, poisoning should only be used in extraordinary situations and efforts must be made to avoid harming nontargeted species.

A seemingly more humane way to handle nuisance raccoons is to move or translocate them, which is a rather common practice in some areas. In Illinois, a large number are moved to rural woodlots or forest preserves. In 1994, more than 5,800 were translocated in this state, a figure that appears to be representative of the numbers moved in recent years. Though moving wildlife to rural areas would seem to be both a logical and merciful way to handle problem animals, this is not as uncomplicated as it might appear. First of all, a raccoon relocated to the country could still end up near a home, and the problem would just be transferred to a different place. In addition, moving animals might expedite the spread of diseases. The translocation of disease-carrying raccoons has been implicated in the rapid spread of rabies along the eastern United States. Furthermore, translocated raccoons might compete with resident ones for resources and disrupt a population's social structure. They could also elevate the predation level on certain species, such as various songbirds, which may already be severe in some areas. Finally, the survival of translocated raccoons is not well understood. Some studies reveal that more than half die within a few months of their release, whereas others indicate that survival rates can be high.

Throughout North America, probably few if any places exist in which raccoons could be successfully released that do not already have sizable populations of humans, raccoons, or both. Most of the natural habitats that are close to urban or suburban areas are small and cannot accommodate many nuisance animals. In some areas, *euthanasia* (a euphemism for killing) may be an alternative to moving problem animals, but of course this is a controversial solution. Perhaps sterilization or the use of reproductive inhibitors could help in dealing with problem animals, but such activities are labor intensive, costly, and questionable in their effectiveness. In determining the best policy for dealing with nuisance raccoons, the consequences of translocation must be weighed against those of such other strategies.

It could appear from this discussion that the only ways in which people are involved with raccoons are through hunting and trapping, or in the removal of those that are offensive. Undoubtedly, these are our principal interactions with raccoons and thus their management strategies focus on these activities. Yet this animal has contributed to our lives in other ways that have important historical, societal, and cultural considerations. Such aspects of the raccoon's role in human affairs are explored in the final chapter, "Raccoons and Humans."

11

Raccoons and Humans

T he raccoon has long held our fascination, admiration, and even our disdain. Perhaps no other animal has been as associated with the history of the United States as has this one. The raccoon has been and continues to be such an integral part of the nation's culture that Whitney and Underwood in *The Coon Hunter's Handbook* contended that it should be designated as our national animal. As society has changed, so has our perception of the raccoon and its place in our lives. The connection between raccoons and humans varies from one geographic area to another, and may also differ across socioeconomic groups. Yet overall, the masked bandit still evokes strong responses because of its unique role in the culture and economy of so many regions.

Given their historical abundance in North America, many Native American tribes probably hunted the raccoon. Evidence for them has been found in the refuse of Native Americans dating from 6000 to 4000 B.C. in what is now the Middle Atlantic United States. Their use has been documented for later inhabitants of the Ohio Valley between A.D. 1000 to 1700, and for more recent tribes such as the Delaware, Sauk, and Potawatomi. Although some tribes reportedly considered its meat to be unappetizing, the raccoon was probably a common source of food in many areas given its ubiquity and the numerous ways of preparing it.

Furthermore, the raccoon was very meaningful to certain tribes. For example, "Raccoon" was one of several "name groups" of both the Kickapoo and Shawnee tribes. In her informative book *Raccoons in Folklore, History and Today's Backyards,* Virginia Holmgren recounts several legends and the many different names Native Americans had for this animal. First of all, various tribes used terms that referred to its manual dexterity. The one the Cree and the Chippewas used means "they pick up things," while that of the Lenape Delaware is translated as "they use their hands as a tool." Other names referred to its body parts, such as the face, forepaws, or tail; its eating of crabs or crayfish; and intriguingly, its perceived magical powers. Some Sioux tribes, for example, bestowed names on the raccoon that seem to have been derived from terms meaning "one who is sacred" or "one with magic." These commonly appear phonetically as *wee-kah, wee-kah-sah,* or *wee-chah,* though they may alternately begin with an "m." A Dakota Sioux tribe reportedly called it *wee-kah teg-alega,* which means "sacred one with painted face."

The Aztec term for the raccoon was *mapachitli* or *mapachitl,* which translates as "one who takes everything in his hands." Their term for the female, however, was *see-oh-at-la-ma-kas-kay,* which means "she who talks with gods." Furthermore, their term for a female with cubs, *ee-yah-mah-tohn,* translates as "little old one who knows things." This is the title that they conferred upon the "wise women" in their society, demonstrating the elevated status of mother raccoons in Aztec culture. Raccoon idolization also seems to have occurred near the Columbia River in the Pacific Northwest. In this area, a Native American carving of a giant face is purported to be the land's guardian spirit called *tsa ga gla lal* (or *tsagiglalal*), meaning "she who watches." Although the Yakima tribe now refers to this figure as the "witch woman" or "watcher," it clearly resembles a raccoon.

Many tribes believed that the raccoon was either related to the dog or that it had doglike qualities. The Tupi of Brazil placed it in a group with dogs and other animals that use a crouching-springing attack to pounce on their prey. Each of these was called a *uara* or "leaper," a word that was modified to reflect a type of leaping. The raccoon was an *aguara-popay,* a "doglike leaper upon crabs and crayfish." The Taino of the Bahamas used the word *ah-on* for raccoons as well as dogs. Though they kept both as pets, they also ate them. The Klamath of the Pacific Northwest also kept the raccoon as a pet, referring to it as a *wacgina,* an "animal that can be tamed like a dog." Yet evidence does not support the notion that raccoons and dogs are closely related. As discussed in Chapter 2, the dog family appeared as a distinct entity considerably before the raccoon family emerged about 25 million years ago.

Given its pervasive association with American culture, it is fitting that we can attribute the first written record of the raccoon, as well as its first European-language name, to the celebrated explorer Christopher Columbus. Soon after landing in the Caribbean in 1492, he noticed a couple of pets near a fisherman's hut on what he named Fernandina Island (now Long Island). In his journals, Columbus referred to them as *perros*, apparently presuming that they were dogs or doglike animals. Perhaps in reference to its mask, he also used the term *perro mastin* or "clownlike dog." Some of his crew members called it a *perro tejon*, a "badgerlike dog." Columbus brought several back to Europe and had them exhibited in Portugal, where they apparently did not survive for long. During his second voyage, raccoon meat was a staple food item for his sailors. In fact, overhunting led to the raccoon's first local extinction on Española, the island where he had established his headquarters. Later known as Hispaniola, today this island includes the Dominican Republic and Haiti.

In the 1520s, a group of Spaniards ventured north from the Caribbean, exploring the islands off of what are now Georgia and South Carolina. Upon returning to their post, they informed its historian that they had observed foxlike animals with faces that were *"muy pintada,"* indicating that they were "painted." This suggests that they had seen raccoons, and if so, it would have been the first European report of them in what is now the United States.

In his 1570s study of Mexican animals, the Spanish physician-naturalist Francisco Hernández indicated that the term for "badgerdog" was the name that the New World Spaniards used most often for the raccoon. But after learning that the Aztecs called it a *mapachitli*, he is said to have prodded the colonists into adopting this term. Eventually, it evolved into *mapache*, which now refers to the common and crab-eating raccoons of Central and South America. Names such as *mapach* and *tepe Maxtlaton* have also been recorded, and *mapachín* is presently used in Costa Rica. Not surprisingly, Spanish-speaking countries have different colloquial names for the raccoon, such as the aforementioned *osito lavador* or "little bear-washer" and *tejon solitaria* or "solitary badger" in Mexico. The latter is reminiscent of the other erroneous associations of raccoons with badgers. By the mid-seventeenth century, raccoons were probably observed in southern Mexico and were seen on the Trés Marías Islands off of western Mexico in 1686.

The earliest reference to the raccoon in Western literature is by another historical figure, Captain John Smith, who is linked to the famous Native American princess, Pocahontas. He reportedly first spelled it as *rahaughcums* as early as 1608. Undoubtedly, Smith was attempting to transcribe the vocalizations of the Algonquins who lived near Virginia's Jamestown colony. As mentioned in Chapter 1, they may have pronounced the word *arakunem* (which translates as "he who scratches with his hands") as *"ah-rah-koon-em."* "Raccoon" is derived from *arakun,* the shortened version of *arakunem.* In 1612, Smith used another spelling for it: "There is a beast they call Aroughcun, much like a badger, but vseth [*sic*] to live on trees as Squirrels doe [*sic*]." He also spelled its plural form as *aroucouns* and *raro-cuns.* Several years ago, archaeologists found a raccoon skull from the seventeenth century near what might be Jamestown Fort on Virginia's Jamestown Island. The teeth are so worn that it probably lived for an unusually long time, perhaps as a pet. Though the main character in the popular 1995 Disney movie *Pocahontas* did indeed have a raccoon pal named Meeko, the real princess is just as likely to have had one for dinner.

By the seventeenth century, various tribes in the Great Lakes region were apparently involved in the fur trade. While the beaver was its primary focus, raccoon pelts were also traded. The area near the southern shore of Lake Erie, the southernmost Great Lake, was close to the northern limit of the raccoon's distribution in this part of the continent. Thus, the region's northern tribes initially might not have recognized the raccoon pelts offered by those from its southern reaches. A northern Huron tribe may have been the first there to use the term "big-tailed" for both the pelts and the tribes bearing them, and the other local tribes evidently adopted this label. These southern fur traders hence became known as "people of the long-tailed ones" and the nearby lake was called "lake of the long-tailed ones." The French settlers in this area transcribed the Huron terms for these as *iri* or *eri,* often adding the suffix *a-chis* (which meant "of this kind"), thus producing the name *irabachis.* The English colonists spelled this as *ee-ree-a-gee,* which then again was shortened to *erie.* This is why the southernmost Great Lake is named Erie, as was the tribe of fur traders from its southern shore.

In 1612, the famed explorer Samuel de Champlain had a map drawn of Lake Erie with a sketch of a raccoon at its northern corner. It included the first written record of the term *iribachis* to identify both the lake and its namesake animal. Curiously though, the area's French settlers apparently used the word *iri* to describe a wild cat, perhaps the bobcat, and labeled the Erie people as *les Chats* and *la nation de Chat.* (Conversely, the Canadian French term *chat sauvage,* though defined as "wild cat," may have referred to the raccoon; a seventeenth-century description of the area's "wild cats" appears to depict one.) It seems highly doubtful though that the Huron words *iri* or *eri* originally referred to a cat. In all likelihood, both the lake and the tribe were named after the raccoon. After all, the Erie's distinctive fur blankets that had tails around their edges were undoubtedly made from raccoon pelts. Yet even today, many of the people in this area still seem to assume that Erie refers to a mountain lion. This confusion could have originated with the French mistaking what the natives called an *iri* for a wild cat, or possibly with the English settlers who thought that the French were referring to a cat.

In any event, of all of the North American tribes, the raccoon has been most closely associated with the Erie. The name actually refers to several tribes centered in northwestern New York south of Lake Erie who seem to have been culturally and linguistically connected to various Northern Iroquoian people. This categorization, however, is based on seventeenth-century French accounts; no Erie words have ever been recorded. There is, at best, an incomplete understanding of this group because of a lack of recorded contact with the European settlers. By the mid-seventeenth century, they had dispersed from the lake area and ultimately disappeared.

Other early descriptions of the raccoon include those of John Lawson and Mark Catesby from the first half of the eighteenth century, who both reported its occurrence in the Carolinas. During that period, Hans Sloane stated that it was common in the mountains of Jamaica, which is odd given that no specimens have been found on this island. In the 1730s, Nathaniel Bailey's English dictionary and other works listed the "raccoon" as a New England animal, whereas the "rattoon" was described as a comparable form in the West Indies. Undoubtedly, these were simply different names for the common raccoon and the various island species.

Just as it is logical to assume that Native Americans would have made use of raccoons wherever they were abundant, the early European settlers in North America must have also used them to satisfy various needs. Their fur was probably converted into coats, cushions, covers, and, of course, the ubiquitous coonskin cap. The size, shape, and thickness of the raccoon's pelt made it a common choice for a winter hat, as evidenced by the numerous illustrations of frontiersmen wearing them. Even before the Revolutionary War, it was recognized as a symbol of the frontier, due in part to the popularity of the legendary backwoodsman Daniel Boone (1734–1820). Benjamin Franklin himself is said to have donned one rather than wear the traditional wig while on his 1776 journey to France.

There is an amusing tale about the raccoon hat from the Revolutionary War. Financial support for the war effort was often lacking, so much so that the Continental soldiers could not count on having adequate uniforms. Consequently, some resorted to wearing their coonskin caps in the winter months. In New Jersey, several units had so many men wearing them that they were supposedly teased about enlisting raccoons. They were greeted with shouts of "Oh, you Raccoons!"

Ironically, the person most identified with the coonskin cap, Davy Crockett (1786–1836), may not have even worn one. He is not wearing one in several portraits that were made while he was alive, which were reproduced in a recent biography by Mark Derr. Instead, he is portrayed in various other caps, including some made from other furbearers. Yet in a portrait made more than 50 years after his death, he is shown holding a coonskin cap at his side. The portrayal of Crockett in a raccoon hat may thus be based on mythology rather than reality. In fact, the characterization of frontiersmen in buckskin and coonskin caps is part of the folklore of that period. Even today, the trustworthy *Encyclopedia Britannica* refers to Crockett as a "coonskin" politician, perhaps implying that he was a backwoodsman. Driven by Walt Disney's marketing genius, this imagery was used to generate the first fad based on a television hero: the wearing of a Davy Crockett coonskin cap. According to the Disney studios, the cap was "a national symbol" that carried enormous power. The craze is worth reviewing, both as part of this discussion of the raccoon's role in U.S. history as well as to describe its effects on their populations.

The fad began after the broadcast of "Davy Crockett, Indian Fighter" in December 1954. It was the first episode of a trilogy that was then released the following May as the movie *Davy Crockett, King of the Wild Frontier*. Another irony is that the cap worn by Fess Parker, the actor who played Davy, was not even made from a raccoon. He alternates between wearing one resembling a combination of badger and coyote with one that appears to be all coyote. Hence, it is plausible that neither the real Crockett nor the actor who portrayed him wore a coonskin cap. Still, this fad generated the greatest demand for raccoon pelts since the coat craze of the 1920s, so much so that the demand for their fur quickly exceeded its supply. Pelt prices jumped from $.25 to $6.00 per pound and the tail rose similarly in value. Raccoon populations were decimated, causing trappers to seek wolves, foxes, skunks, and opossums to meet the need for fur. Caps were also made from recycled coats, including those from the previous fad. One line of caps produced from cardboard and crepe paper was found to be flammable and fire chiefs across the country were alerted to their danger. During this period, Davy Crockett products accounted for 10 percent of all of the children's clothing sales in the United States.

In addition to using the raccoon for clothing, the early European settlers probably considered it to be an acceptable type of meat. Like other game, raccoons were presumably roasted and strips of their meat were smoked like bacon. Their fat was used for many purposes: it could be applied as a salve for bruises and sprains, it was converted into a lubricant and employed as a leather softener, and it was probably used in place of lard. The raccoon became so highly valued that the number of pelts a man

A young boy reads about American pioneer legend Davy Crockett while wearing a coonskin cap and a Davy Crockett T-shirt. Photographed in 1955 in Topeka, Kansas. Copyright Bettmann/CORBIS.

owned was regarded as a measure of his success. Its fur was so popular in Europe that the pelts provided the early settlers with a steady source of income. Indeed, high quality pelts could be substituted for money at trading posts and were used to pay court fines. In the state of Franklin, which became part of Tennessee, low-ranking government officials received their salaries in coonskins.

The raccoon's significance in the country's early days is reflected in many phrases and terms, as well as its various namesakes. The expression "a coon's age" was used by settlers to suggest a long time, though it is not known if this was based on a belief about its longevity. By the 1830s, other sayings invoking the raccoon also became common. "Old-timers," especially trappers, referred to themselves as "this yer old coon," which might have been a way of boasting about their survival skills. A person walking over a log bridge on all fours was said to have "cooned" their way across, and one "cooned" for melons by furtively eating them in the field. The yellow perch became known as the "raccoon perch" because fishermen believed that raccoons readily preyed upon them. For the same reason, oysters on the rocks closest to the shore were called "raccoon oysters." Wild grapes were "raccoon grapes" and various wild berries were known as "raccoon berries." Many towns, creeks, rivers, and mountains are named both after the raccoon and variants of its Native American names. For example, the Sioux term for it, *wee-kah* or *wee-chah*, which means "little man," is the basis for the names of such cities as Wichita and Ouachita.

The masked bandit even figured prominently in one of the nation's early presidential elections. The 1840 race was between President Martin Van Buren, the Democratic Party nominee, and William Henry Harrison, the Whig Party candidate. Van Buren was viewed by many, especially in the rural areas, as a "citified" type. Alternately, Harrison, who was honored as an Indian fighter at the battle of Tippecanoe in Indiana and later in the War of 1812, appealed to the nation's less cosmopolitan sector. The former conflict was the basis of the campaign slogan, "Tippecanoe and Tyler too," used by him and John Tyler, his running mate. Harrison's 1840 run for the White House, his second, is considered to be the nation's first "packaged" presidential candidacy. Despite the fact that he was actually a member of the country's aristocracy, he was marketed as a simple backwoods soul. The Whig symbol during this election was a log cabin with a cider jug by the door and a coonskin tacked above it. Henry Clay, a founder of the Whig Party and Speaker of the House of Representatives, was called "Old Coon." However, Harrison also became known as "Old Coon," and he easily won the election and became the country's ninth president. His nickname, however, continued to be used by his adversaries as a source of ridicule. His partisans' speeches were lampooned as having been made in the "Coongress," and they were derided as being loaded with "coonisms." Unfortunately, Harrison died after only a month in office; the cold that he caught during his lengthy inaugural speech had developed into pneumonia.

Some 80 years after the Old Coon briefly occupied the White House, two real raccoons lived there: A female named Rebecca lived with Calvin and Grace Coolidge, and one called Suzie later joined Herbert and Lou Hoover. Paradoxically, the raccoon coat became the height of fashion during this era, the Roaring Twenties. With about 15 skins used per coat, their numbers were depleted in various areas. Fortunately, the fad ended before their populations dwindled to dire levels. Wildlife officials, though, decided to release individuals in some places to augment their numbers. This was probably the first and one of the only times that a major effort was undertaken to restore diminished raccoon populations.

The raccoon has been especially connected with the culture of the southeastern United States and is featured prominently in the region's poetry, stories, and songs. Many of these tales involve interactions between raccoon and opossums. In his insightful book *Southern Hunting in Black and White*, Stuart Marks comments that the opossum has a lower status than the raccoon in the rural South. Their difference in rank is reflected in various works such as "Shake the Persimmons Down," which was published in Thomas Washington Talley's 1922 volume *Negro Folk Rhymes*:

> *De raccoon up in de 'simmon tree,*
> *Dat 'possum on de groun',*
> *De 'possum say to de raccoon:*
> *"Suh!"*
> *"Please shake dem 'simmons down."*
> *De raccoon say to de 'possum: "Suh!"*
> *(As he grin from down below),*
> *"If you wants dese good 'simmons, man,*
> *Jes clam up whar dey grow."*

Marks also provides a fascinating look at the culture of raccoon hunting in south-central North Carolina. He contrasts the hunting of raccoons with that of foxes. Whereas a fox hunt is usually an extravagant daytime event conducted by wealthy landowners, raccoon hunting is a simpler affair that normally occurs at night and has long been identified as an activity of the poor, slaves, and then later freed persons who needed the food. One of the reasons that raccoon hunting may appeal to those with little wealth is because few dogs are needed for it. Thus, although the hunts may no longer be driven by a need for food, they remain attractive to poor rural residents. African Americans, many of whom remain economically disadvantaged in the rural South, reportedly account for a disproportionately large segment of raccoon consumers.

Even though those who hunt them may be stigmatized as poor people, raccoon hunting is still a widespread activity. In South Carolina, for example, it has grown significantly in popularity over the last 20 years. The United Kennel Club licenses more than 6,000 coonhound events nationally each year, a number that has increased in recent times. At these competitions, the taking of game is not allowed. Coonhounds play a central role in raccoon hunting and clubs are typically organized around them. As is the case for other kinds of hunting, the dogs used to pursue raccoons are specialized. An understanding of their training, as well as their buying and selling, is essential for appreciating the culture of raccoon hunting.

At various competitions, the unregistered dogs of the poor often challenge the registered ones of more affluent hunters. Some of the participants at these field trials may disparage those who can afford dogs with particular pedigrees. Nevertheless, the use of registered coonhounds is common and the number of breeds may be a reflection of the sport's popularity. The six breeds used are: the Black and Tan, Blue Tick, English, Plott, Redbone, and (Treeing) Walker hounds. Most of these qualify for United Kennel Club certification as a breed, but only the Black and Tan is recognized by the group many consider to be the premier dog organization in the United States, the American Kennel Club. Several popular magazines are also devoted to coonhounds, such as *Full Cry* and *Coonhound Bloodlines.* Coonhound events are designed to test a dog's ability to perform under hunting conditions. They are held by organizations such as the United Kennel Club and the National Coon Dog Field Trial Association, which sponsors the highly regarded Leafy Oak Trials. There is even a Coonhunting World Championship. Without question, though, it is the dog's performance during the hunt that ultimately matters most.

Coonhounds were like many other commodities that were brought to North America from Europe and then adapted for use in their new circumstances. Given that many of the potential game animals such as the raccoon find refuge in trees, the early settlers would have clearly benefited from having dogs that could scare animals into trees or at least locate them once they fled there. Consequently, breeds with these qualities were produced, supposedly from English foxhounds, German boxers, red Irish hounds, Cuban bloodhounds, and perhaps other strains.

George Washington might be the father of coonhound breeding as well as the father of his country. He may have been the first North American to possess the Grand Bleu de Gascogne hounds (French Blue Gascon hounds) from which some of today's coonhounds are derived. In 1785, he received seven of them from the Marquis de Lafayette of France. Much admired for their keen noses and tireless efforts, these dogs have the dubious distinction of driving France's wolves to extinction in the Middle Ages. For centuries, they were bred to pursue animals that remained on the ground, such as hares and wild boar. Thus, four months after he acquired them, Washington grumbled that he was "plagued with the Dogs running Hogs." Ultimately, these hounds or their descendants must have developed a treeing capability, as they became used to hunt game that retreated to the trees.

By the early 1800s, several new breeds had emerged that eventually showed great success in pursuing raccoons. Simon Kent, a scout in the Ohio Valley, helped breed the first Black and Tan coonhounds, possibly from the Gascon hounds and a strain known as the Kerry beagle. In the bayous of Louisiana, the descendants of French trappers produced Blue Tick coonhounds, which also strongly resemble the Gascon hounds. It is not known if the animals that gave rise to either the Black and Tan or the Blue Tick coonhounds were descended from Washington's dogs. Another coonhound, the Plott, originated in the Great Smoky Mountains. The descendants of a German immigrant, Johannes Plott, first used dogs called *schweisshunds* to produce a brindle-patterned hound that may have originally been used to hunt bears.

Raccoon hunting is usually learned early in life from one's family or neighbors. Marks maintains that its teaching can affect a rural Southerner's identity and that this activity can even take on a transcendental quality for such individuals. In an unusual twist on the notion that the hunt can be spiritually uplifting, an Athens, Georgia, congregation raised more than $2,000 by staging a "Coon Hunt for Christ" in 1995. Its reverend submitted that "The coon hunt is a way to spread the word of God, to talk about Jesus Christ." Hunting raccoons has functioned in other colorful ways: It has served as a cover for hiding illegal whiskey production, womanizing, and acts of violence.

The raccoon has continued to be represented in our entertainment culture in numerous ways. Another 1950s television show that incorporated the raccoon was *The Honeymooners*. It featured a bus driver named Ralph Kramden, played by Jackie Gleason, who belonged to a fraternal order called the Grand Exalted Brotherhood of Raccoons. In the next decade, at least two popular rock-and-roll songs made reference to the raccoon: Bob Dylan's hit "Subterranean Homesick Blues" included a verse about a man who wore a coonskin cap, and one of the Beatles's later songs, "Rocky Raccoon," recounted the escapades of a character by that name. Several years ago, the Renault company considered producing an all-terrain vehicle called the Raccoon; its logo depicted the masked bandit's face. As mentioned earlier, the 1995 movie *Pocahontas* had a raccoon named Meeko that was marketed in various forms such as stuffed toys. Marvel Comics, the company that spawned Spiderman and other well-known characters, even created a raccoon superhero, Rocket Raccoon. Today, raccoons are commonly featured in children's books and television shows, and there is a cartoon production company called Raccoon Network. In perhaps one of the oddest twists, Fess Parker, the actor who portrayed Davy Crockett, now sells coonskin-covered bottle tops at his California winery.

Naturally, the raccoon has become thoroughly modern; it has several Web sites on the Internet. One of the best is "The World Wide Raccoon Web" (*www.loomcom.com/raccoons*). It contains a considerable amount of useful and interesting material, including such sections as "Raccoon Facts and Information" and "The Raccoon Gallery" (a collection of images and sounds), as well as links to other raccoon Web sites.

The raccoon has clearly been an intriguing part of our history. Its on-going success in North America ensures that it will remain a fascinating part of our culture for many generations to come.

LIST OF SCIENTIFIC NAMES

These are the scientific names for the common names of the fungi, plants, and animals—other than the raccoon and its family—referred to in the text. The term *likely* is used when the correctness of the scientific name is in question. The abbreviation spp. for species is used to indicate that the common name shown applies to several species of the given genus, whereas sp. is used when only the genus name of the plant or animal mentioned in the text is known. Individual names are not provided for those groups that are composed of various genera, such as lizards or oysters.

Fungi
Chestnut blight fungus—*Endothia parasitica*

Plants
Trees and Shrubs
　Apple—*Malus* sp.
　Beech—*Fagus grandifolia*
　Chestnut, American—*Castanea dentata*
　Citrus—*Citrus* spp.
　Cherry—*Prunus* sp.
　Figs—*Ficus* sp. and *Ficus carica*
　Gum, tupelo—likely *Nyssa aquatica*
　Hackberry—*Celtis* sp.

Hickory—*Carya* sp.
Juniper—*Juniperus* sp.
Mangrove, American or red—*Rhizophora mangle*
Manzanita—*Arctostaphylos* sp.
Maple, red—*Acer rubrum*
Marlberry—*Ardisia* sp.
Mesquite—*Prosopis glandulosa*
Oak—*Quercus* spp.
Oak, live—*Quercus virginiana*
Oak, swamp chestnut—*Quercus prinus*
Olive, Russian—*Elaeagnus angustifolia*
Palmetto—likely *Sabal palmetto*
Peach—*Prunus persica*
Pecan—*Carya illinoensis*
Persimmon—*Diospyros virginiana*

Pine—*Pinus* spp.
Plum—*Prunus* sp.
Walnut—*Juglans* sp.
Other
Barley—likely *Hordeum vulgare*
Corn—*Zea mays*
Grape—*Vitis* sp.
Juneberry—*Amelanchier* sp.
Millet, foxtail—*Setaria italica*
Raspberry—*Rubus* spp.
Rush—*Juncus* sp.
Salt-marsh grass; Cordgrass—
Spartina sp.
Sorghum—*Sorghum vulgare*
Spanish moss—*Tillandsia usneoides*
Sunflower—*Helianthus* sp.
Oats—*Avena sativa*
Watermelon—*Citrullus vulgaris*
Wheat—*Triticum* sp.

Invertebrates
Crab, fiddler—*Uca* sp.
Crayfish—*Cambarus* sp. and
Astacus sp.
Earthworm—*Lumbricus terrestris*
Grasshopper—*Melanoplus* sp.

Fishes
Bass, largemouth—*Micropterus salmoides*
Bluegill—*Lepomis macrochirus*
Bullhead—*Ictalurus* sp.
Catfish—*Ictalurus* sp.
Carp—*Cyprinus carpio*
Eel, American—*Anguilla rostrata*
Gar—*Lepisosteus* sp.
Perch, yellow—*Perca flavescens*
Pickerel—*Esox* sp.
Pike—*Esox* sp.

Shad—likely *Alosa* sp.
Sucker—likely *Catostomus* sp.
Sunfish—*Lepomis* sp.
Trout—likely *Oncorhyncus* sp. and
Salvelinus sp.

Amphibians
Salamander, tiger—*Ambystoma tigrinum*
Treefrog, gray—*Hyla versicolor*

Reptiles
Alligator—*Alligator mississippiensis*
Snakes
Garter snake—*Thamnophis* sp.
Water snake—*Natrix* sp.
Turtles
Cooter—*Pseudemys* sp.
Green turtle—*Chelonia mydas*
Hawksbill turtle—*Eretmochelys imbricata*
Leatherback turtle—*Dermochelys coriacea*
Loggerhead turtle—*Caretta caretta*
Mud turtle—*Kinosternon* sp.
Slider—*Trachemys* sp.
Snapping turtle—likely
Chelydra serpentina
Softshell turtle—*Trionyx* sp.

Birds
Blackbird, red-winged—*Agelaius phoeniceus*
Blackbird, yellow-headed—
Xanthocephalus xanthocephalus
Cormorant, double-crested—
Phalacrocorax auritus
Coot, American—*Fulica americana*

Cowbird, brown-headed—
Molothrus ater
Duck, canvasback—*Aythya
valisineria*
Duck, mallard—*Anas platyrhynchos*
Duck, wood—*Aix sponsa*
Grackle, common—*Quiscalus
quiscula*
Gull, herring—*Larus argentatus*
Hawk, red-tailed—*Buteo jamaicensis*
Heron, black-crowned night—
Nycticorax nycticorax
Heron, great blue—*Ardea herodias*
Magpie—*Pica* sp.
Moorhen, common—*Gallinula
chloropus*
Murrelet, ancient—
Synthliboramphus antiquus
Owl, great horned—*Bubo
virginianus*
Owl, short-eared—*Asio flammeus*
Pheasant, ring-necked—*Phasianus
colchicus*
Quail, bobwhite—*Colinus
virginianus*
Rail—likely *Rallus* sp.
Turkey—*Meleagris gallopavo*

Mammals
Aardwolf—*Proteles cristatus*
Armadillo, nine-banded—*Dasypus
novemcinctus*
Anteater, spiny; Echidna
Australia—*Tachyglossus aculeatus*
New Guinea—*Zaglossus bruijni*
Badger—*Taxidea taxus*
Bandicoot—*Perameles* sp.
Bear, black—*Ursus americanus*
Bear, grizzly—*Ursus arctos*
Beaver—*Castor canadensis*

Beaver (from Miocene epoch)—
Paleocastor sp.
Binturong—*Arctictis binturong*
Boar, wild—*Sus scrofa*
Bobcat—*Lynx rufus*
Cat—*Felis catus*
Cougar—*Puma concolor*
Cow—*Bos taurus*
Coyote—*Canis latrans*
Deer, musk—*Moschus moschiferus*
Deer, white-tailed—*Odocoileus
virginianus*
Dog—*Canis domesticus*
Elephant, African (savanna)—
Loxodonta africana
Fisher—*Martes pennanti*
Fox, red—*Vulpes vulpes*
Fox, gray—*Urocyon cinereoargenteus*
Genet—*Genetta* spp.
Gopher—likely *Geomys* spp.
Hare; Jackrabbit—*Lepus* spp.
Horse—*Equus caballus*
Hyena, spotted—*Crocuta crocuta*
Jaguar—*Panthera onca*
Kangaroo—*Macropus* spp.
Lynx, Canada—*Lynx canadensis*
Marten—*Martes americana*
Mink—*Mustela vison*
Mole, likely eastern—*Scalopus
aquaticus*
Moose—*Alces alces*
Mouse—*Peromyscus* spp.
Mouse, house—*Mus musculus*
Muskrat—*Ondatra zibethicus*
Noolbenger; Honey possum—
Tarsipes rostratus
Opossum, Virginia—*Didelphis
virginiana*
Panda, giant—*Ailuropoda
melanoleuca*
Platypus—*Ornithorhyncus anatinus*

Rabbit, Cottontail—*Sylvilagus* spp.

Rat, Norway—*Rattus norvegicus*

Sable—*Martes zibellina*

Shrew—likely *Sorex* spp.

Skunk, striped—*Mephitis mephitis*

Squirrel, eastern fox—*Sciurus niger*

Squirrel, eastern gray—*Sciurus carolinensis*

Squirrel, ground—*Spermophilus* spp.

Vole, meadow—*Microtus pennsylvanicus*

Weasel—*Mustela* spp.

Wolf, gray—*Canis lupus*

Woodchuck—*Marmota monax*

R E F E R E N C E S

Preface
Rue, L. L. 1964. The World of the Raccoon. Philadelphia: Lippincott.
Zeveloff, S. I. 1997. Significant issues in the evolution of the Procyonidae. Abstracts of Oral and Poster Papers, Seventh International Theriological Congress, Acapulco, Mexico: 395.

Chapter 1
Holmgren, V. C. 1990. Raccoons in Folklore, History and Today's Backyards. Santa Barbara, Calif.: Capra Press.
Lyall-Watson, M. 1963. A critical re-examination of food "washing" behaviour in the raccoon (*Procyon lotor* Linn.). Proceedings of the Zoological Society of London 141:371–393.
Radinsky, L. 1976. Cerebral clues. Natural History 85 (5): 54–59.
Rue, L. L. 1964. The World of the Raccoon. Philadelphia: Lippincott.

Chapter 2
Bailey, B., and R. M. Hunt, Jr. 1997. The oldest New World procyonids: Origins, geologic setting, and biogeography. Abstracts of Oral and Poster Papers, Seventh International Theriological Congress, Acapulco, Mexico: 31.
Baskin, J. A. 1982. Tertiary Procyoninae (Mammalia: Carnivora) of North America. Journal of Vertebrate Paleontology 2:71–93.
———. 1989. Comments on New World Tertiary Procyonidae (Mammalia, Carnivora). Journal of Vertebrate Paleontology 9:110–117.
———. 1997. The fossil record of the New World Procyonidae. Abstracts of Oral and Poster Papers, Seventh International Theriological Congress, Acapulco, Mexico: 40–41.
———. 1998a. Procyonidae. Pp. 144–151. *In* C. M. Janis, K. M. Scott, and L. L. Jacobs (eds.), Evolution of Tertiary Mammals of North America. Vol. I: Terrestrial Carnivores, Ungulates, and Ungulatelike Mammals. Cambridge, U.K.: Cambridge University Press.
———. 1998b. Mustelidae. Pp. 152–173. *In* C. M. Janis, K. M. Scott, and L. L. Jacobs (eds.), Evolution of Tertiary Mammals of North America. Vol. I: Terrestrial Carnivores, Ungulates, and Ungulatelike Mammals. Cambridge, U.K.: Cambridge University Press.
Baskin, J. A., and R. H. Tedford. 1996. Small arctoid and feliform carnivorans. Pp. 486–497. *In* D. R. Prothero and R. J. Emry (eds.), The Terrestrial Eocene-

Oligocene Transition in North America. Cambridge, U.K.: Cambridge University Press.

Berta, A. 1988. Quaternary evolution and biogeography of the large South American Canidae (Mammalia: Carnivora). University of California Publications in Geological Sciences 132:1–149.

Bininda-Emonds, O. R. P., J. L. Gittleman, and A. Purvis. 1999. Building large trees combining phylogenetic information: A complete phylogeny of the extant Carnivora (Mammalia). Biological Reviews 74:143–175.

Decker, D. M., and W. C. Wozencraft. 1991. Phylogenetic analysis of Recent procyonid genera. Journal of Mammalogy 72:42–55.

Eisenberg, J. F. 1981. The Mammalian Radiations: An Analysis of Trends in Evolution, Adaptation, and Behavior. Chicago: University of Chicago Press.

———. 1989. An Introduction to the Carnivora. Pp. 1–9. In J. L. Gittleman (ed.), Carnivore Behavior, Ecology, and Evolution. Vol. 1. Ithaca, N.Y.: Comstock Publishing Associates, Cornell University Press.

Ginsburg, L. 1961. La faune des carnivores Miocènes de Sansan (Gers). Mémoires du Musée National Histoire Naturelle, Paris Nouvelle Série, Série C, 9:1–190.

Honacki, J. H., K. E. Kinman, and J. W. Koeppl. 1982. Mammal Species of the World: A Taxonomic and Geographic Reference. Lawrence, Kans.: Allen Press and Association of Systematics Collections.

Hunt, R. M., Jr. 1996. Biogeography of the Order Carnivora. Pp. 485–541. In J. L. Gittleman (ed.), Carnivore Behavior, Ecology, and Evolution. Vol. 2. Ithaca, N.Y.: Comstock Publishing Associates, Cornell University Press.

Janis, C. M., J. A. Baskin, A. Berta, J. J. Flynn, G. F. Gunnell, R. M. Hunt Jr., L. D. Martin, and K. Munthe. 1998. Carnivorous mammals. Pp. 73–90. In C. M. Janis, K. M. Scott, and L. L. Jacobs (eds.), Evolution of Tertiary Mammals of North America. Vol. I: Terrestrial Carnivores, Ungulates, and Ungulatelike Mammals. Cambridge, U.K.: Cambridge University Press.

Macdonald, D. 1992. The Velvet Claw: A Natural History of the Carnivores. London: BBC Books.

Marshall, L. G. 1985. Geochronology and land mammal biochronology of the transamerican faunal interchange. Pp. 49–85. In F. Stehli and S. D. Webb (eds.), The Great American Biotic Interchange. New York: Plenum Press.

Martin, L. D. 1989. Fossil history of the terrestrial Carnivora. Pp. 536–568. In J. L. Gittleman (ed.), Carnivore Behavior, Ecology, and Evolution. Vol. 1. Ithaca, N.Y.: Comstock Publishing Associates, Cornell University Press.

Mayr, E. 1986. Uncertainty in science: Is the giant panda a bear or a raccoon? Nature 323:769–771.

McKenna, M. C., and S. K. Bell. 1997. Classification of Mammals above the Species Level. New York: Columbia University Press.

O'Brien, S. J. 1987. The ancestry of the giant panda. Scientific American 257 (5):102–107.

O'Brien, S. J., M. Menotti-Raymond, W. J. Murphy, W. G. Nash, J. Wienberg, R. Stanyon, N. G. Copeland, N. A. Jenkins, J. E. Womack, and J. A. Marshall Graves. 1999. The promise of comparative genomics in mammals. Science 286:458–481.

Pecon Slattery, J., and S. J. O'Brien. 1995. Molecular phylogeny of the red panda (*Ailurus fulgens*). Journal of Heredity 86:413–422.

Radinsky, L. B. 1981a. Evolution of skull shape in carnivores. 1. Representative modern carnivores. Biological Journal of the Linnean Society 15:369–388.

———. 1981b. Evolution of skull shape in carnivores. 2. Additional modern carnivores. Biological Journal of the Linnean Society 16:337–355.

Sarich, V. 1973. The giant panda is a bear. Nature 245:218–220.

Schaller, G. B. 1993. The Last Panda. Chicago: University of Chicago Press.

Tedford, R. H. 1976. Relationships of pinnipeds to other carnivores (Mammalia). Systematic Zoology 25:363–374.

———. 1994a. Caught in time. Natural History 103 (4): 90–91.

———. 1994b. Key to the carnivores. Natural History 103 (4): 74–76.

Vaughan, T. A., J. M. Ryan, and N. J. Czaplewski. 2000. Mammalogy. 4th ed. Orlando, Fla.: Saunders College Publishing.

Vrana, P. B., M. C. Milinkovitch, J. R. Powell, and W. C. Wheeler. 1994. Higher level relationships of the Arctoid Carnivora based on sequence data and "total evidence." Molecular Phylogenetics and Evolution 3:47–58.

Wayne, R. K., R. E. Benveniste, D. N. Janczewski, and S. J. O'Brien. 1989. Molecular and biochemical evolution of the Carnivora. Pp. 465–494. In J. L. Gittleman (ed.), Carnivore Behavior, Ecology, and Evolution. Vol. 1. Ithaca, N.Y.: Comstock Publishing Associates, Cornell University Press.

Wolsan, M. 1993. Phylogeny and classification of early European Mustelida (Mammalia: Carnivora). Acta Theriologica 38:345–384.

———. 1997. The oldest procyonids and the primitive procyonid morphology. Abstracts of Oral and Poster Papers, Seventh International Theriological Congress, Acapulco, Mexico: 373–374.

Wolsan, M., and B. Lange-Badré. 1996. An arctomorph carnivoran skull from the Phosphorites du Quercy and the origin of procyonids. Acta Palaeontologica Polonica 41:277–298.

Woodburne, M. O., ed. 1987. Cenozoic Mammals of North America: Geochronology and Biostratigraphy. Berkeley: University of California Press.

Wozencraft, W. C. 1989a. The phylogeny of the Recent Carnivora. Pp. 495–535. In J. L. Gittleman (ed.), Carnivore Behavior, Ecology, and Evolution. Vol. 1. Ithaca, N.Y.: Comstock Publishing Associates, Cornell University Press.

———. 1989b. Classification of the Recent Carnivora. Pp. 569–593. In J. L. Gittleman (ed.), Carnivore Behavior, Ecology, and Evolution. Vol. 1. Ithaca, N.Y.: Comstock Publishing Associates, Cornell University Press.

Wozencraft, W. C., and R. S. Hoffman. 1993. How the bears came to be. Pp. 14–22. In I. Stirling (ed.), Bears: Majestic Creatures of the Wild. Emmaus, Pa.: Rodale Press.

Wurster, D. H., and K. Benirschke. 1968. Comparative cytogenetic studies in the Order Carnivora. Chromosoma 24:336–382.

Wyss, A. R., and J. J. Flynn. 1993. A phylogenetic analysis and definition of the Carnivora. Pp. 32–52. In F. S. Szalay, M. J. Novacek, and M. C. McKenna (eds.), Mammal Phylogeny: Placentals. New York: Springer-Verlag.

Zeveloff, S. I. 1997. Significant issues in the evolution of the Procyonidae. Abstracts of Oral and Poster Papers, Seventh International Theriological Congress, Acapulco, Mexico: 395.

Chapter 3

Baskin, J. A. 1982. Tertiary Procyoninae (Mammalia: Carnivora) of North America. Journal of Vertebrate Paleontology 2:71–93.

———. 1998. Procyonidae. Pp. 144–151. In C. M. Janis, K. M. Scott, and L. L. Jacobs (eds.), Evolution of Tertiary Mammals of North America. Vol. I: Terrestrial Carnivores, Ungulates, and Ungulatelike Mammals. Cambridge, U.K.: Cambridge University Press.

Bonaparte, C. L. J. L. 1845. Catalogo methodico dei mammiferi Europei. Milan: L. di Giacomo Pirola.

CITES. Web site of the Convention on International Trade in Endangered Species of Wild Fauna and Flora. [Cited on September 15, 2001.] Available from *http://www.cites.org.*

Decker, D. M., and W. C. Wozencraft. 1991. Phylogenetic analysis of Recent procyonid genera. Journal of Mammalogy 72:42–55.

De La Rosa, C. L., and C. D. Nocke. 2000. A Guide to the Carnivores of Central America: Natural History, Ecology, and Conservation. Austin: University of Texas Press.

Eisenberg, J. F. 1981. The Mammalian Radiations: An Analysis of Trends in Evolution, Adaptation, and Behavior. Chicago: University of Chicago Press.

———. 1989. Mammals of the Neotropics. Vol. 1: The Northern Neotropics. Chicago: University of Chicago Press.

Emmons, L. H. 1990. Neotropical Rainforest Mammals: A Field Guide. Chicago: University of Chicago Press.

Gilbert, B. 2000. Ringtails like to be appreciated. Smithsonian 31 (5): 64–70.

Glatston, A. R., compiler. 1994. The Red Panda, Olingos, Coatis, Raccoons, and Their Relatives: Status Survey and Conservation Action Plan for Procyonids and Ailurids. Gland, Switzerland: International Union for the Conservation of Nature and Natural Resources.

Gompper, M. E. 1995. *Nasua narica.* Mammalian Species No. 487. American Society of Mammalogists.

Gompper, M. E., and D. M. Decker. 1998. *Nasua nasua.* Mammalian Species No. 580. American Society of Mammalogists.

Gray, J. E. 1825. Outline of an attempt at the disposition of the Mammalia into tribes and families with a list of the genera apparently appertaining to each tribe. Annals of Philosophy, n.s., 10:337–344.

Hunter, M. L., Jr. 1991. Conservation strategies for giant and red pandas. Trends in Ecology and Evolution 6:379–380.

McKenna, M. C., and S. K. Bell. 1997. Classification of Mammals Above the Species Level. New York: Columbia University Press.

Morgan, G. S., C. E. Ray, and O. Arredondo. 1980. A giant extinct insectivore from Cuba (Mammalia: Insectivora: Solenodontidae). Proceedings of the Biological Society of Washington 93:597–608.

Morgan, G. S., and C. A. Woods. 1986. Extinction and the zoogeography of West Indian land mammals. Biological Journal of the Linnean Society 28:167–203.

Pecon Slattery, J., and S. J. O'Brien. 1995. Molecular phylogeny of the red panda (*Ailurus fulgens*). Journal of Heredity 86:413–422.

Pons, J.-M., V. Volobouev, J.-F. Ducroz, A. Tillier, and D. Reudet. 1999. Is the Guadeloupean racoon (*Procyon minor*) really an endemic species? New insights from molecular and chromosomal analyses. Journal of Zoological Systematics and Evolutionary Research 37:101–108.

Radinsky, L. B. 1981. Evolution of skull shape in carnivores. 2. Additional modern carnivores. Biological Journal of the Linnean Society 16:337–355.

Stone, D. 1995. Raccoons and their Relatives. Cambridge, U.K.: International Union for the Conservation of Nature.

Wolsan, M. 1993. Phylogeny and classification of early European Mustelida (Mammalia: Carnivora). Acta Theriologica 38:345–384.

———. 1997. The oldest procyonids and the primitive procyonid morphology. Abstracts of Oral and Poster Papers, Seventh International Theriological Congress, Acapulco, Mexico: 373–374.

World Conservation Union. The 2000 IUCN Red List of Threatened Species. [Cited September 30, 2000.] Available from *http://www.redlist.org.*

Wozencraft, W. C. 1989a. The phylogeny of the Recent Carnivora. Pp. 495–535. *In* J. L. Gittleman (ed.), Carnivore Behavior, Ecology, and Evolution. Vol. 1. Ithaca, N.Y.: Comstock Publishing Associates, Cornell University Press.

———. 1989b. Classification of the Recent Carnivora. Pp. 569–593. *In* J. L. Gittleman (ed.), Carnivore Behavior, Ecology, and Evolution. Vol. 1. Ithaca, N.Y.: Comstock Publishing Associates, Cornell University Press.

Chapter 4

Bergmann, C. 1847. Ueber die verhältnisse der warmeökonomie der theire zu ihrer grösse. Gottinger Studien. 3 (1): 595–708.

Boyce, M. S. 1979. Seasonality and patterns of natural selection for life histories. American Naturalist 114:569–583.

Cole, L. W. 1907. Concerning the intelligence of raccoons. Journal of Comparative Neurology and Psychology 17:211–262.

———. 1912. Observations of the senses and instincts of the raccoon. Journal of Animal Behaviour 2:299–309.

Gehrt, S. D., and E. K. Fritzell. 1999. Growth rates and intraspecific variation in body weights of raccoons (*Procyon lotor*) in southern Texas. American Midland Naturalist 141:19–27.

Hall, E. R. 1981. The Mammals of North America. Vol. II. 2d ed. New York: John Wiley and Sons.

Ritke, M. E., and M. L. Kennedy. 1988. Intraspecific morphologic variation in the raccoon *Procyon lotor* and its relationship to selected environmental variables. Southwestern Naturalist 33:295–314.

Sanderson, G. C. 1987. Raccoon. Pp. 487–499. *In* M. Novak, J. A. Baker, M. E. Obbard, and B. Malloch (eds.), Wild Furbearer Management and Conservation in North America. Ontario, Canada: Ministry of Natural Resources.

Schaller, G. B. 1993. The Last Panda. Chicago: University of Chicago Press.

Shedd, W. 2000. Owls Aren't Wise and Bats Aren't Blind: A Naturalist Debunks Our Favorite Fallacies about Wildlife. New York: Harmony Books.

Sieber, O. J. 1986. Acoustic recognition between mother and cubs in raccoons (*Procyon lotor*). Behaviour 96:130–163.

Stains, H. J. 1957. A clue to spring molt in raccoons as revealed in analysis of scats. Illinois Academy of Science Transactions 50:257–258.

Stuewer, F. W. 1943. Raccoons: Their habits and management in Michigan. Ecological Monographs 13:203–257.

Whitney, L. F., and A. B. Underwood. 1952. The Coon Hunter's Handbook. New York: Henry Holt and Company.

Zeveloff, S. I., and M. S. Boyce. 1988. Body size patterns in North American mammal faunas. *In* M. S. Boyce (ed.), Evolution of Life Histories of Mammals: Theory and Pattern. New Haven, Conn.: Yale University Press.

Chapter 5

Aliev, F. F., and G. C. Sanderson. 1966. Distribution and status of the raccoon in the Soviet Union. Journal of Wildlife Management 30:497–502.

Bailey, V. 1926. A biological survey of North Dakota. North American Fauna 49. Washington, D.C.: Bureau of Biological Survey, U.S. Department of Agriculture.

Burris, O. E., and D. E. McKnight. 1973. Game transplants in Alaska. Wildlife Technical Bulletin 4. Juneau: Alaska Department of Fish and Game.

de Beaufort, F. 1968. Apparition de raton-laveur, *Procyon lotor* (L.) en France. Mammalia 32:307.

Finley, R. B., Jr. 1995. The spread of raccoons (*Procyon lotor hirtus*) into the Colorado Plateau from the Northern Great Plains. Proceedings of the Denver Museum of Natural History, Series 3, No. 11.

Glatston, A. R., compiler. 1994. The Red Panda, Olingos, Coatis, Raccoons, and Their Relatives: Status Survey and Conservation Action Plan for Procyonids and Ailurids. Gland, Switzerland: International Union for the Conservation of Nature and Natural Resources.

Goldman, E. A. 1950. Raccoons of North and Middle America. North American Fauna 60. Washington, D.C.: U.S. Fish and Wildlife Service, U.S. Department of the Interior.

Greenwood, R. J., and M. A. Sovada. 1999. Population trends for prairie pothole carnivores. Pp. 461–463. *In* M. J. Mac, P. A. Opler, C. E. Puckett Haecker, and P. D. Doran (eds.), Status and Trends of the Nation's Biological Resources. Vol. 2. Washington, D.C.: U.S. Geological Survey, U.S. Department of the Interior.

Hart, E. B. 1982. The raccoon, *Procyon lotor*, in Wyoming. Great Basin Naturalist 42:599–600.

Porter, W. F., and J. A. Hill. 1999. Northeast. Pp. 181–218. *In* M. J. Mac, P. A. Opler, C. E. Puckett Haecker, and P. D. Doran (eds.), Status and Trends of the Nation's Biological Resources. Vol. 1. Washington, D.C.: U.S. Geological Survey, U.S. Department of the Interior.

Redford, P. 1962. Raccoon in the U.S.S.R. Journal of Mammalogy 43:541–542.

Sanderson, G. C. 1951. Breeding habits and a history of the Missouri raccoon population from 1941 to 1948. Transactions of the North American Wildlife Conference 16:445–461.

Scheffer, V. B. 1947. Raccoons transplanted in Alaska. Journal of Wildlife Management 11:350–351.

White, A. B. 1986. Food habits of a Salt Lake Valley raccoon (*Procyon lotor*) population. Unpublished paper for non-thesis master's degree, University of Utah, Salt Lake City.

White, S., P. K. Kennedy, and M. L. Kennedy. 1998. Temporal genetic variation in the raccoon, *Procyon lotor*. Journal of Mammalogy 79:747–754.

Yalden, D. 1999. The History of British Mammals. London: T & AD Poyser.

Chapter 6

Berner, A., and L. W. Gysel. 1967. Raccoon use of large tree cavities and ground burrows. Journal of Wildlife Management 31:706–714.

Bider, J. R., P. Thibault, and R. Sarrazin. 1968. Schèmes dynamiques spatio temporels de l'activité de *Procyon lotor* en relation avec le comportement. Mammalia 32:137–163.

Butterfield, R. T. 1944. Populations, hunting pressure, and movement of Ohio raccoons. Transactions of the North American Wildlife Conference 9:337–344.

———. 1954. Traps, live-trapping, and marking of raccoons. Journal of Mammalogy 35:440–442.

Cabalka, J. L., R. R. Costa, and G. O. Hendrickson. 1953. Ecology of the raccoon in central Iowa. Proceedings of the Iowa Academy of Science 60:616–620.

Cowan, W. F. 1973. Ecology and life history of the raccoon (*Procyon lotor hirtus* Nelson and Goldman) in the northern part of its range. Ph.D. diss., University of North Dakota, Grand Forks.

Dasmann, R. F. 1981. Wildlife Biology. New York: John Wiley and Sons.

Dorney, R. S. 1954. Ecology of marsh raccoons. Journal of Wildlife Management 18:217–225.

Fritzell, E. K. 1978. Habitat use by prairie raccoons during the waterfowl breeding season. Journal of Wildlife Management 42:118–127.

———. 1982. Comments on raccoon population eruptions. Wildlife Society Bulletin 10:70–72.

Fritzell, E. K., and R. J. Greenwood. 1984. Mortality of raccoons in North Dakota. Prairie Naturalist 16:1–4.

Gehrt, S. D., L. B. Fox, and D. L. Spencer. 1993. Locations of raccoons during flooding in eastern Kansas. Southwestern Naturalist 38:404–406.

Gehrt, S. D., and E. K. Fritzell. 1997. Sexual differences in home ranges of raccoons. Journal of Mammalogy 78:921–931.

———. 1998a. Duration of familial bonds and dispersal patterns for raccoons in south Texas. Journal of Mammalogy 79:859–872.

———. 1999b. Growth rates and intraspecific variation in body weights of raccoons (*Procyon lotor*) in southern Texas. American Midland Naturalist 141:19–27.

Giles, L. W. 1940. Fall food habits of the raccoon in eastern Iowa. Journal of Wildlife Management 4:375–382.

Glatston, A. R., compiler. 1994. The Red Panda, Olingos, Coatis, Raccoons, and Their Relatives: Status Survey and Conservation Action Plan for Procyonids and Ailurids. Gland, Switzerland: International Union for the Conservation of Nature and Natural Resources.

Glueck, T. F., W. R. Clark, and R. D. Andrews. 1988. Raccoon movement and habitat use during the fur harvest season. Wildlife Society Bulletin 16:6–11.

Greenwood, R. J. 1981. Foods of prairie raccoons during the waterfowl nesting season. Journal of Wildlife Management 45:754–760.

———. 1982. Nocturnal activity and foraging of prairie raccoons (*Procyon lotor*) in North Dakota. American Midland Naturalist 107:238–243.

Hartman, L. H., A. J. Gaston, and D. S. Eastman. 1997. Raccoon predation on ancient murrelets on East Limestone Island, British Columbia. Journal of Wildlife Management 61:377–388.

Hoffman, C. O., and J. L. Gottschang. 1977. Numbers, distribution, and movements of a raccoon population in a suburban residential community. Journal of Mammalogy 58:623–636.

Ivey, D. W. R. 1948. The raccoon in the salt marshes of northeastern Florida. Journal of Mammalogy 29:290–291.

Jobin, B., and J. Picman. 1977. Factors affecting predation on artificial nests in marshes. Journal of Wildlife Management 61:792–800.

Kadlec, J. 1971. Effects of introducing foxes and raccoons on herring gull colonies. Journal of Wildlife Management 35:625–636.

Lehman, L. E. 1977. Population ecology of the raccoon on the Jasper-Pulaski Wildlife Study Area. Pittman-Robertson Bulletin 9. Indianapolis, Ind.: Division of Fisheries and Wildlife, Indiana Department of Natural Resources.

———. 1984. Raccoon density, home range, and habitat use on south-central Indiana farmland. Pittman-Robertson Bulletin 15. Indianapolis, Ind.: Division of Fisheries and Wildlife, Indiana Department of Natural Resources.

Llewellyn, L. M., and C. G. Webster. 1960. Raccoon predation on waterfowl. Transactions of the North American Wildlife and Natural Resources Conference 25:180–185.

Lynch, G. M. 1967. Long-range movement of a raccoon in Manitoba. Journal of Mammalogy 48:659–660.

Maier, T. J., and R. M. DeGraaff. 2000. Predation on Japanese quail vs. house sparrow eggs in artificial nests: Small eggs reveal small predators. Condor 102:325–332.

McLaughlin, J. H. 1953. Factors influencing the raccoon and its management in southwestern Virginia. Master's thesis, Virginia Polytechnic Institute, Blacksburg.

Mech, L. D., D. M. Barnes, and J. R. Tester. 1968. Seasonal weight changes, mortality, and population structure of raccoons in Minnesota. Journal of Mammalogy 54:496–498.

Moore, J. C. 1953. Raccoon parade. Everglades Natural History 1:119–126.

Mugaas, J. N., J. Seidensticker, and K. P. Mahlke-Johnson. 1993. Metabolic adaptation to climate and distribution of the raccoon *Procyon lotor* and other Procyonidae. Smithsonian Contributions to Zoology, No. 542. Washington D.C.: Smithsonian Institution Press.

Nottingham, B. G., K. G. Johnson, J. W. Woods, and M. R. Pelton. 1982. Population characteristics and harvest relationships of a raccoon population in east Tennessee. Proceedings of the Annual Conference of Southeastern Association of Game and Fish Agencies 36:691–700.

Pedlar, J. H., L. Fahrig, and H. G. Merriam. 1997. Raccoon habitat use at 2 spatial scales. Journal of Wildlife Management 61:102–112.

Porter, W. F., and J. A. Hill. 1999. Northeast. Pp. 181–218. *In* M. J. Mac, P. A. Opler, C. E. Puckett Haecker, and P. D. Doran (eds.), Status and Trends of the Nation's Biological Resources. Vol. 1. Washington, D.C.: U.S. Geological Survey, U.S. Department of the Interior.

Priewert, F. W. 1961. Record of an extensive movement by a raccoon. Journal of Mammalogy 42:113.

Ratnaswamy, M. J., R. J. Warren, M. T. Kramer, and M. D. Adam. 1997. Comparisons of lethal and nonlethal techniques to reduce raccoon depredation of sea turtle nests. Journal of Wildlife Management 61:368–376.

Rivest, P., and J. M. Bergeron. 1981. Density, food habits, and economic importance of raccoons (*Procyon lotor*) in Quebec agrosystems. Canadian Journal of Zoology 59:1755–1762.

Salt Lake Tribune. 1997. Raccoon boom is bad for Utah birds. April 16:D4.

Samson, F. B., F. L. Knopf, and W. R. Ostlie. 1999. Grasslands. Pp. 437–472. *In* M. J. Mac, P. A. Opler, C. E. Puckett Haecker, and P. D. Doran (eds.), Status and Trends of the Nation's Biological Resources. Vol. 2. Washington, D.C.: U.S. Geological Survey, U.S. Department of the Interior.

Sanderson, G. C. 1951. Breeding habits and a history of the Missouri raccoon population from 1941 to 1948. Transactions of the North American Wildlife Conference 16:445–461.

Schneider, D. G., L. D. Mech, and J. R. Tester. 1971. Movements of female raccoons and their young as determined by radio-tracking. Animal Behaviour 41:1–43.

Schoonover, L. J., and W. H. Marshall. 1951. Food habits of the raccoon (*Procyon lotor hirtus*) in north-central Minnesota. Journal of Mammalogy 32:422–428.

Seton, E. T. 1909. The Arctic Prairies. New York: Charles Scribner's Sons.

Sharp, W. M., and L. H. Sharp. 1956. Nocturnal movements and behavior of wild raccoons at a winter feeding station. Journal of Mammalogy 37:170–177.

Sonenshine, D. E., and E. L. Winslow. 1972. Contrasts in distribution of raccoons in two Virginia localities. Journal of Wildlife Management 36:838–847.

Stains, H. J. 1961. Comparison of temperatures inside and outside two tree dens used by raccoons. Ecology 42:410–413.

Stoudt, J. H. 1971. Ecological factors affecting waterfowl production in the Saskatchewan Parklands. Resource Publication 99. Washington, D.C.: U.S. Fish and Wildlife Service, U.S. Department of the Interior.

Stuewer, F. W. 1943. Raccoons: Their habits and management in Michigan. Ecological Monographs 13:203–257.

———. 1948. Artificial dens for raccoons. Journal of Wildlife Management 12:296–301.

Twichell, A. R., and H. H. Dill. 1949. One hundred raccoons from one hundred and two acres. Journal of Mammalogy 30:130–133.

Tyson, E. L. 1950. Summer food habits of the raccoon in southwest Washington. Journal of Mammalogy 31:448–449.

Urban, D. 1970. Raccoon populations, movement patterns, and predation on a managed waterfowl marsh. Journal of Wildlife Management 34:372–382.

White, A. B. 1986. Food habits of a Salt Lake Valley raccoon (*Procyon lotor*) population. Unpublished paper for non-thesis master's degree, University of Utah, Salt Lake City.

Whitney, L. F., and A. B. Underwood. 1952. The Coon Hunter's Handbook. New York: Henry Holt and Company.

Wood, J. E. 1954. Food habits of furbearers of the upland post oak region in Texas. Journal of Mammalogy 35:406–415.

Yeager, L. E. 1937. Naturally sustained yield in a farm fur crop in Mississippi. Journal of Wildlife Management 1:28–36.

———. 1943. Fur production and management of Illinois drainage system. Transactions of the North American Wildlife Conference 8:294–301.

Yeager, L. E., and W. H. Elder. 1945. Pre- and post-hunting season foods of raccoons on an Illinois goose refuge. Journal of Wildlife Management 9:48–56.

Yeager, L. E., and R. G. Rennels. 1943. Fur yield and autumn foods of the raccoon in Illinois river bottom lands. Journal of Wildlife Management 7:45–60.

Zeveloff, S. I., and P. D. Doerr. 1985. Seasonal weight changes in raccoons (Carnivora: Procyonidae) of North Carolina. Brimleyana 11:63–67.

Chapter 7

Chamberlain, M. J., K. M. Hodges, B. D. Leopold, and T. S. Wilson. 1999. Survival and cause-specific mortality of adult raccoons in central Mississippi. Journal of Wildlife Management 63:880–888.

Clark, W. R., J. J. Hasbrouck, J. M. Kienzler, and T. F. Glueck. 1989. Vital statistics and harvest of an Iowa raccoon population. Journal of Wildlife Management 53:982–990.

Fritzell, E. K., and R. J. Greenwood. 1984. Mortality of raccoons in North Dakota. Prairie Naturalist 16:1–4.

Gehrt, S. D., and E. K. Fritzell. 1999. Survivorship of a nonharvested raccoon population in south Texas. Journal of Wildlife Management 63:889–894.

Glueck, T. F., W. R. Clark, and R. D. Andrews. 1988. Raccoon movement and habitat use during the fur harvest season. Wildlife Society Bulletin 16:6–11.

Hasbrouck, J. J., W. R. Clark, and R. D. Andrews. 1992. Factors associated with raccoon mortality in Iowa. Journal of Wildlife Management 56:693–699.

U.S. Geological Survey. Raccoon rabies: Example of translocation, disease. [Cited September 6, 2000.] Available from *http://biology.usgs.gov.*

Llewellyn, L. M., and C. G. Webster. 1960. Raccoon predation on waterfowl. Transactions of the North American Wildlife and Natural Resources Conference 25:180–185.

Mech, L. D., D. M. Barnes, and J. R. Tester. 1968. Seasonal weight changes, mortality, and population structure of raccoons in Minnesota. Journal of Mammalogy 54:496–498.

Milius, S. 1998. No raccoon boom after vaccination program. Science News 153:277.

Minser, W. G. III, and M. R. Pelton. 1982. The impact of hunting on raccoon populations and management implications. Bulletin 612. Knoxville: University of Tennessee Agricultural Experiment Station.

Nottingham, B. G., K. G. Johnson, J. W. Woods, and M. R. Pelton. 1982. Population characteristics and harvest relationships of a raccoon population in east Tennessee. Proceedings of the Annual Conference of Southeastern Association of Game and Fish Agencies 36:691–700.

Rosatte, R. C., P. M. Kelly-Ward, and C. D. MacInnes. 1987. A strategy for controlling rabies in urban skunks and raccoons. Pp. 54–60. *In* L. W. Adams and D. L. Leedy (eds.), Integrating Man and Nature in the Metropolitan Environment. Columbia, Md.: Proceedings of the National Symposium on Urban Wildlife. National Institute for Urban Wildlife.

Shedd, W. 2000. Owls Aren't Wise and Bats Aren't Blind: A Naturalist Debunks Our Favorite Fallacies about Wildlife. New York: Harmony Books.

Shoop, C. R., and C. A. Ruckdeschel. 1990. Alligators as predators on terrestrial mammals. American Midland Naturalist 124:407–412.

Whitney, L. F., and A. B. Underwood. 1952. The Coon Hunter's Handbook. New York: Henry Holt and Company.

Winkler, W. G., and K. Bögel. 1992. Control of rabies in wildlife. Scientific American 266 (6): 86–92.

Chapter 8

Bissonette, T. H., and A. G. Csech. 1938. Sexual photoperiodicity of raccoons on low protein diet and second litters in the same breeding season. Journal of Mammalogy 19:342–348.

Clark, W. R., J. J. Hasbrouck, J. M. Kienzler, and T. F. Glueck. 1989. Vital statistics and harvest of an Iowa raccoon population. Journal of Wildlife Management 53:982–990.

Fritzell, E. K. 1978. Reproduction of raccoons (*Procyon lotor*) in North Dakota. American Midland Naturalist 100:253–256.

Fritzell, E. K., G. F. Hubert Jr., B. E. Meyen, and G. C. Sanderson. 1985. Age-specific reproduction in Illinois and Missouri raccoons. Journal of Wildlife Management 49:901–905.

Gehrt, S. D., and E. K. Fritzell. 1996. Second estrus and late litters in raccoons. Journal of Mammalogy 77:388–393.

———. 1998. Duration of familial bonds and dispersal patterns for raccoons in south Texas. Journal of Mammalogy 79:859–872.

———. 1999. Growth rates and intraspecific variation in body weights of raccoons (*Procyon lotor*) in southern Texas. American Midland Naturalist 141:19–27.

McLaughlin, I. H. 1953. Factors influencing the raccoon and its management in southwestern Virginia. Master's thesis, Virginia Polytechnic Institute, Blacksburg.

Nottingham, B. G., K. G. Johnson, J. W. Woods, and M. R. Pelton. 1982. Population characteristics and harvest relationships of a raccoon population in east Tennessee. Proceedings of the Annual Conference of Southeastern Association of Game and Fish Agencies 36:691–700.

Sanderson, G. C., and A. V. Nalbandov. 1973. The reproductive cycle of the raccoon in Illinois. Illinois Natural History Survey Bulletin 31 (2): 29–85.

Zeveloff, S. I., and P. D. Doerr. 1981. Reproduction of raccoons in North Carolina. Journal of the Elisha Mitchell Scientific Society 97:194–199.

Chapter 9

Darwin, C. 1859. On the Origin of Species by Means of Natural Selection, or the Preservation of Favoured Races in the Struggle for Life. London: J. Murray.

———. 1871. The Descent of Man, and Selection in Relation to Sex. London: J. Murray.

Darwin, C., and A. Wallace. 1858. On the Tendency of Species to Form Varieties; and on the Perpetuation of Varieties and Species by Natural Means of Selection. Proceedings [of the Linnean Society of London] 3:45–62.

Ellis, R. J. 1964. Tracking raccoons by radio. Journal of Wildlife Management 28:363–368.

Fritzell, E. K. 1978. Aspects of raccoon (*Procyon lotor*) social organization. Canadian Journal of Zoology 56:260–271.

Gehrt, S. D., and E. K. Fritzell. 1997. Sexual differences in home ranges of raccoons. Journal of Mammalogy 78:921–931.

———. 1998a. Duration of familial bonds and dispersal patterns for raccoons in south Texas. Journal of Mammalogy 79:859–872.

———. 1998b. Resource distribution, female home range dispersion and male spatial interactions: Group structure in a solitary carnivore. Animal Behaviour 55:1211–1227.

———. 1999. Behavioural aspects of the raccoon mating system: Determinants of consortship success. Animal Behaviour 57:593–601.

Glueck, T. F., W. R. Clark, and R. D. Andrews. 1988. Raccoon movement and habitat use during the fur harvest season. Wildlife Society Bulletin 16:6–11.

Hoffman, C. O., and J. L. Gottschang. 1977. Numbers, distribution, and movements of a raccoon population in a suburban residential community. Journal of Mammalogy 58:623–636.

Kaufmann, J. H. 1971. Is territoriality definable? Pp. 34–60. *In* A. H. Esser (ed.), Behavior and Environment: The Use of Space by Animals and Men. New York: Plenum Press.

Lehman, L. E. 1977. Population ecology of the raccoon on the Jasper-Pulaski Wildlife Study Area. Pittman-Robertson Bulletin 9. Indianapolis, Ind.: Division of Fisheries and Wildlife, Indiana Department of Natural Resources.

Mech, L. D., and F. J. Turkowski. 1966. Twenty-three raccoons in one winter den. Journal of Mammalogy 47:529–530.

Porterfield, T. R. 1980. Status and management investigations of the raccoon in the mountains of North Carolina. Final Report, Project Number W-57. Raleigh: North Carolina Wildlife Resources Commission.

Ritke, M. E. 1990. Sexual dimorphism in the raccoon (*Procyon lotor*): Morphological evidence for intrasexual selection. American Midland Naturalist 124:342–351.

Stuewer, F. W. 1943. Raccoons: Their habits and management in Michigan. Ecological Monographs 13:203–257.

Chapter 10

Amundson, R. 1971. The raccoon. Wildlife in North Carolina 35 (3): 10–11.

Chamberlain, M. J., K. M. Hodges, B. D. Leopold, and T. S. Wilson. 1999. Survival and cause-specific mortality of adult raccoons in central Mississippi. Journal of Wildlife Management 63:880–888.

Clark, W. R. 1990. Compensation in furbearer populations: Current data compared with a review of concepts. Transactions of the North American Wildlife and Natural Resources Conference 55:491–500.

Clark, W. R., and E. K. Fritzell. 1992. A review of population dynamics of furbearers. Pp. 899–910. *In* D. R. McCullough and R. H. Barrett (eds.), Wildlife 2001: Populations. New York: Elsevier Applied Science.

Clark, W. R., J. J. Hasbrouck, J. M. Kienzler, and T. F. Glueck. 1989. Vital statistics and harvest of an Iowa raccoon population. Journal of Wildlife Management 53:982–990.

Conner, M. C., and R. F. Labisky. 1985. Evaluation of radioisotope tagging for estimating abundance of raccoon populations. Journal of Wildlife Management 49:326–332.

Conner, M. C., R. F. Labisky, and D. R. Progulske Jr. 1983. Scent-station indices as measures of population abundance for bobcats, raccoons, gray foxes, and opossums. Wildlife Society Bulletin 11:146–152.

Edwards, T. L. 1985. Occurrence of officer-reported raccoon hunting law violators in Kentucky. Proceedings of the Annual Conference of Southeastern Association of Fish and Wildlife Agencies 39:540–547.

Fritzell, E. K., G. F. Hubert Jr., B. E. Meyen, and G. C. Sanderson. 1985. Age-specific reproduction in Illinois and Missouri raccoons. Journal of Wildlife Management 49:901–905.

Gaines, K. F., C. G. Lord, C. S. Boring, I. L. Brisbin Jr., M. Gochfeld, and J. Burger. 2000. Raccoons as potential vectors of radionuclide contamination to human food chains from a nuclear industrial site. Journal of Wildlife Management 64:199–208.

Gehrt, S. D., and E. K. Fritzell. 1996. Second estrus and late litters in raccoons. Journal of Mammalogy 77:388–393.

———. 1999. Survivorship of a nonharvested raccoon population in south Texas. Journal of Wildlife Management 63:889–894.

Hodges, K. M., M. J. Chamberlain, and B. D. Leopold. 2000. Effects of summer hunting on ranging behavior of adult raccoons in central Mississippi. Journal of Wildlife Management 64:194–198.

International Association of Fish and Wildlife Agencies. n.d. U.S. fur harvest (1970–1995) and fur value (1974–1995) statistics by state and region. Washington, D.C.

Khan, A. T., S. J. Thompson, and H. W. Mielke. 1995. Lead and mercury levels in raccoons from Macon County, Alabama. Bulletin of Environmental Contamination and Toxicology 54:812–816.

Kramer, M. T., R. J. Warren, M. J. Ratnaswamy, and B. T. Bond. 1999. Determining sexual maturity of raccoons by external versus internal aging criteria. Wildlife Society Bulletin 27:231–234.

Lord, R. D., Jr. 1960. Litter size and latitude in North American mammals. American Midland Naturalist 64:488–499.

Minser, W. G., III, and M. R. Pelton. 1982. The impact of hunting on raccoon populations and management implications. Bulletin 612. Knoxville: University of Tennessee Agricultural Experiment Station.

Moses, R. A., and S. Boutin. 1987. Aging raccoons in Ontario by logistic regression on pelt sizes. Journal of Wildlife Management 51:820–824.

Mosillo, M., E. J. Heske, and J. D. Thompson. 1999. Survival and movements of translocated raccoons in northcentral Illinois. Journal of Wildlife Management 63:278–286.

Nottingham, B. G., K. G. Johnson, J. W. Woods, and M. R. Pelton. 1982. Population characteristics and harvest relationships of a raccoon population in east Tennessee. Proceedings of the Annual Conference of Southeastern Association of Game and Fish Agencies 36:691–700.

Novak, M., M. E. Obbard, J. G. Jones, R. Newman, A. Booth, A. J. Satterthwaite, and G. Linscombe. 1987. Furbearer Harvests in North America, 1600–1984. (Supplement to Wild Furbearer Management and Conservation in North America). Ontario, Canada: Ministry of Natural Resources.

Rosatte, R. C., and C. D. MacInnes. 1989. Relocation of city raccoons. Pp. 87–92. *In* A. D. Bjugstad, D. W. Uresk, and R. H. Hamre (technical coordinators), U.S. Department of Agriculture Forest Service General Technical Report RM-171. Ninth Great Plains Wildlife Damage Control Workshop Proceedings, Fort Collins, Colo.

Sanderson, G. C. 1950. Methods of measuring productivity in raccoons. Journal of Wildlife Management 14:389–402.

Sanderson, G. C., and G. F. Hubert, Jr. 1981. Selected demographic characteristics of Illinois (U.S.A.) raccoons (*Procyon lotor*). Pp. 487–513. *In* J. A. Chapman and D. Pursley (eds.), Worldwide Furbearer Conference Proceedings, Vol. I, Frostburg, Md.

Semel, B., and L. K. Nicolaus. 1992. Estrogen-based aversion to eggs among free-ranging raccoons. Ecological Applications 2:439–449.

Seton, E. T. 1909. The Arctic Prairies. New York: Charles Scribner's Sons.

———. 1929. The Lives of Game Animals. Vol. 2. Garden City, N.Y.: Doubleday.

Stuewer, F. W. 1943. Raccoons: Their habits and management in Michigan. Ecological Monographs 13:203–257.

———. 1948. Artificial dens for raccoons. Journal of Wildlife Management 12:296–301.

Stone, D. 1995. Raccoons and Their Relatives. Cambridge, U.K.: International Union for the Conservation of Nature.

White, S., P. K. Kennedy, and M. L. Kennedy. 1998. Temporal genetic variation in the raccoon, *Procyon lotor*. Journal of Mammalogy 79:747–754.

Wilson, K. A. 1955. Fur resources of North Carolina. Pittman-Robertson Project W-6-R. Raleigh: North Carolina Wildlife Resources Commission.

Chapter 11

Bilger, B. 1996. Barking up the right tree. The Atlantic Monthly (January) 277 (1): 94–98.

Callender, C. 1978a. Sauk. Pp. 648–655. *In* B. G. Trigger (ed.), Northeast, Handbook of North American Indians. Vol. 15. W. C. Sturtevant (general ed.), Washington, D.C.: Smithsonian Institution.

———. 1978b. Shawnee. Pp. 622–635. *In* B. G. Trigger (ed.), Northeast, Handbook of North American Indians. Vol. 15. W. C. Sturtevant (general ed.), Washington, D.C.: Smithsonian Institution.

Callender, C., R. K. Pope, and S. M. Pope. 1978. Kickapoo. Pp. 656–667. *In* B. G. Trigger (ed.), Northeast, Handbook of North American Indians. Vol. 15. W. C. Sturtevant (general ed.), Washington, D.C.: Smithsonian Institution.

Clifton, J. A. 1978. Potawatomi. Pp. 725–742. *In* B. G. Trigger (ed.), Northeast, Handbook of North American Indians. Vol. 15. W. C. Sturtevant (general ed.), Washington, D.C.: Smithsonian Institution.

Derr, M. 1993. The Frontiersman: The Real Life and Many Legends of Davy Crockett. New York: William Morrow and Company.

Encyclopedia Britannica. Davy Crockett. [Cited March 2, 2001.] Available from *http://www.britannica.com*.

Funk, R. E. 1978. Post-Pleistocene adaptations. Pp. 16–27. *In* B. G. Trigger (ed.), Northeast, Handbook of North American Indians. Vol. 15. W. C. Sturtevant (general ed.), Washington, D.C.: Smithsonian Institution.

Gaines, K. F., C. G. Lord, C. S. Boring, I. L. Brisbin Jr., M. Gochfeld, and J. Burger. 2000. Raccoons as potential vectors of radionuclide contamination to human food chains from a nuclear industrial site. Journal of Wildlife Management 64:199–208.

Goddard, I. 1978. Delaware. Pp. 213–239. *In* B. G. Trigger (ed.), Northeast, Handbook of North American Indians. Vol. 15. W. C. Sturtevant (general ed.), Washington, D.C.: Smithsonian Institution.

Griffin, J. B. 1978. Late Prehistory of the Ohio Valley. Pp. 547–559. *In* B. G. Trigger (ed.), Northeast, Handbook of North American Indians. Vol. 15. W. C. Sturtevant (general ed.), Washington, D.C.: Smithsonian Institution.

Hartman, C. G. 1952. Possums. Austin: University of Texas Press.

Hassrick, R. B. 1964. The Sioux. Norman: University of Oklahoma Press.

Holmgren, V. C. 1990. Raccoons in Folklore, History and Today's Backyards. Santa Barbara, Calif.: Capra Press.

King, M. J. 1976. The Davy Crockett craze: A case study in popular culture. Ph.D. diss., University of Hawaii, Manoa.

Lind, M. 1998. Opinion: The many Davy Crocketts. New York Times. [Cited on November 23, 1998.] Available from *http://www.nytimes.com*.

Marks, S. A. 1991. Southern Hunting in Black and White: Nature, History, and Ritual in a Carolina Community. Princeton, N.J.: Princeton University Press.

Porterfield, T. R. 1980. Status and management investigations of the raccoon in the mountains of North Carolina. Final Report, Project Number W-57. Raleigh: North Carolina Wildlife Resources Commission.

Sagard-Théodat, G. 1632. Dictionnaire de la langue huronne. Paris: Denys Moreau.

Shepherd, C. 1995. News of the Weird. Salt Lake Tribune, May 21:D5.

Stuewer, F. W. 1943. Raccoons: Their habits and management in Michigan. Ecological Monographs 13:203–257.

Talley, T. W. 1922. Negro Folk Rhymes. New York: Macmillan.

U.S. News and World Report. 1995. Pocahontas, for real. June 19:61–64.

White, M. E. 1978. Erie. Pp. 412–417. *In* B. G. Trigger (ed.), Northeast, Handbook of North American Indians. Vol. 15. W. C. Sturtevant (general ed.), Washington, D.C.: Smithsonian Institution.

White, R. 1991. The Middle Ground, Indians, Empires, and Republics in the Great Lakes Region, 1650–1815. Cambridge, U.K.: Cambridge University Press.

Whitney, L. F., and A. B. Underwood. 1952. The Coon Hunter's Handbook. New York: Henry Holt and Company.

Appendix

Integrated Taxonomic Information System. [Cited August 24, 2001.] Available from *http://www.itis.usda.gov*.

Jones, C., R. S. Hoffmann, D. W. Rice, M. D. Engstrom, R. D. Bradley, D. J. Schmidly, C. A. Jones, and R. J. Baker. 1997. Revised checklist of North American mammals north of Mexico, 1997. Occasional Papers Number 173. Lubbock: Museum of Texas Tech University.

Nowak, R. M. 1999. Walker's Mammals of the World. Vols. I and II. 6th ed. Baltimore, Md.: Johns Hopkins University Press.

PLANTS database. [Cited August 24, 2001.] Available from *http://plants.usda.gov*.

Roca, A. L., N. Georgiadis, J. Pecon-Slattery, and S. J. O'Brien. 2001. Genetic evidence for two species of elephant in Africa. Science 293:1473–1477.

Wilson, D. E., and F. R. Cole. 2000. Common Names of Mammals of the World. Washington, D.C.: Smithsonian Institution Press.

General References

Goldman, E. A. 1950. Raccoons of North and Middle America. North American Fauna 60. Washington, D.C.: U.S. Fish and Wildlife Service, U.S. Department of the Interior.

Hall, E. R. 1981. The Mammals of North America. Vol. II. 2d ed. New York: John Wiley and Sons.

Johnson, A. S. 1970. Biology of the raccoon (*Procyon lotor varius* Nelson and Goldman) in Alabama. Auburn University Agricultural Experiment Station Bulletin 402.

Kaufmann, J. H. 1982. Raccoon and allies. Pp. 567–585. *In* J. A. Chapman and G. A. Feldhammer (eds.), Wild Mammals of North America: Biology, Management, and Economics. Baltimore, Md.: Johns Hopkins University Press.

Lotze, J.-H., and S. Anderson. 1979. *Procyon lotor.* Mammalian Species No. 119. American Society of Mammalogists.

MacClintock, D. 1981. A Natural History of Raccoons. New York: Charles Scribner's Sons.

Macdonald, D. 1992. The Velvet Claw: A Natural History of the Carnivores. London: BBC Books.

Nowak, R. M. 1999. Walker's Mammals of the World. Vols. I and II. 6th ed. Baltimore, Md.: Johns Hopkins University Press.

Sanderson, G. C. 1987. Raccoon. Pp. 487–499. *In* M. Novak, J. A. Baker, M. E. Obbard, and B. Malloch (eds.), Wild Furbearer Management and Conservation in North America. Ontario, Canada: Ministry of Natural Resources.

Whitney, L. F., and A. B. Underwood. 1952. The Raccoon. Orange, Conn.: Practical Science Publishing Company.

INDEX